André Philipp Mundt

Dynamic risk management with Markov decision processes

Dynamic risk management with Markov decision processes

von
André Philipp Mundt

universitätsverlag karlsruhe

Dissertation, Universität Karlsruhe (TH)
Fakultät für Mathematik, 2007
Referenten: Prof. Dr. Nicole Bäuerle, Prof. Dr. Ulrich Rieder

Impressum

Universitätsverlag Karlsruhe
c/o Universitätsbibliothek
Straße am Forum 2
D-76131 Karlsruhe
www.uvka.de

Universitätsverlag Karlsruhe 2008
Print on Demand

ISBN: 978-3-86644-200-9

Preface

During almost four years of work at the Kompetenzzentrum Versicherungswissenschaften GmbH, Leibniz Universität Hannover, and the Institut für Stochastik, Universität Karlsruhe (TH), many people supported me.

First of all, I would like to thank my supervisor Nicole Bäuerle. She draw me to the subject of dynamic risk management and made many fruitful suggestions for the research treated in this work. Furthermore, she always took time for me and encouraged me when I was stuck and came to her with questions and problems. Secondly, let me thank Ulrich Rieder for being the second advisor and for useful discussions.

From all the other people, I firstly and most importantly have to mention Anja Blatter. She read almost the whole manuscript and made many valuable comments that led to improvements in formulations and the layout. Furthermore, I have to thank Gunther Amt and Bruno Ebner for helping me improve parts of the thesis and Mirko Kötter and Lars Michael Hoffmann for useful discussions.

Altogether, my colleagues and former colleagues from the Kompetenzzentrum Versicherungswissenschaften GmbH, the Institut für Mathematische Stochastik in Hanover and the Institut für Stochastik in Karlsruhe always provided a friendly and enjoyable atmosphere at work and during working breaks. When I moved to Hanover and Karlsruhe they made it very easy for me to settle in there.

Finally, I have to mention my family and in particular my parents for supporting my studies and their confidence in my ability to complete my diploma and this thesis.

Karlsruhe,
October 2007

André Mundt

Contents

Introduction

Dealing with *risk* is particularly important for financial corporations such as banks and insurance companies, since these face uncertain events in various lines of their business. There exist several characterizations of risk which do not always have to describe distinct issues. This could be market risk, credit risk or operational risk, but also model risk or liquidity risk. Insurance companies additionally have to deal with underwriting risk, for example. The aim of *integrated risk management* is to manage some or all kinds of potential risk simultaneously. According to McNeil et al. (2005), the more general notion of *risk management* describes a "discipline for living with the possibility that future events may cause adverse effects", while banks or insurance companies are even able to "manage risks by repacking them and transferring them to markets in customized ways". Consequently, risk management can be regarded as a "core competence" of such companies.

Historically, methods of dealing with risk go back to mean–variance optimization problems. These were first considered by de Finetti (1940) (see Barone (2006) for a translated version of the original article in Italian), whereas the issue only became famous with the work of Nobel laureate Markowitz (1952). Before, a main focus was on the mean of the underlying quantities, whereas these works brought the risk component in terms of the variance into play. Currently, some discussion about the contribution of the two authors can be observed in the literature. First, Markowitz (2006) himself claims that de Finetti did not solve the problem for the general case of correlated risk. However, this argument is contrasted in Pressacco and Serafini (2007), where the authors reason that de Finetti solved the problem, but under a regularity assumption which is not always fulfilled.

An important tool in risk management is the implementation of *risk measures*, in particular ones which go beyond the variance. Since the risk is modelled by random quantities, i. e. random variables or, in a dynamic framework, stochastic processes, it is often possible to estimate or approximate the distribution of such positions. However, the actual realization of the risk remains uncertain. One way to quantify the risk beyond its distribution is to use risk measures. Risk measures assign a value to a risk at a future date that can be interpreted as a present monetary value or present and future monetary values if *dynamic risk measures* are applied.

Let us now review some regulatory aspects and describe how risk measures are applied in risk management so far. This aspect started with the first Basel accord and its amendment which introduced the risk measure Value–at–Risk for calculating the regulatory capital in internal models. Furthermore, special emphasis is put on credit and operational risk within the context of Basel II. While these accords aim to stabilize the international banking business, similar regulatory frameworks are introduced for the insurance sector via Solvency 2, having a more local or national character. In particular, the aim is to regulate by law what kind of risk measures should be used to calculate the companies' *target capital* in order to ensure a certain level of solvency. Special emphasis is put on the coherent risk measure *Average Value–at–Risk*, sometimes denoted as *Expected Shortfall*, which is used in Solvency 2 instead of Value–at–Risk in Basel II. It has the advantage of being subadditive for general risks, therefore awarding diversification (see Chapter 1 for a more thorough investigation). Value–at–Risk fulfills this property only in an elliptical world. Consequently, using Value–at–Risk is not always the wrong thing to do but dangerous if applied inappropriately.

Although many problems are similar for the banking and insurance sector respectively, there are some distinctions between these two kinds of firms. Banks mainly deal with bounded risks, e. g. when facing credit risk. On the other hand, insurance companies often have to consider unbounded risks, e. g. when heavy–tailed distributed financial positions are present. To address both situations, we always treat integrable but not necessarily bounded risks in this work. Furthermore, a main issue will be to develop risk management tools for dynamic models. These naturally occur when considering portfolio optimization problems or in the context of developing reasonable risk measures for final payments or even stochastic processes. We consider only models in discrete time and denote these approaches with *dynamic risk management*.

In dynamic economic models, one often faces a Markovian structure of the underlying stochastic processes. Hence, a method we will frequently use is the theory of *Markov decision processes (MDP)*, sometimes also addressed as *dynamic programming (DP)* introduced in Bellman (1952). In this work, one field where the method is applied is dynamic portfolio optimization. This is a standard approach. However, the theory can even be used to solve optimization problems that occur in economically motivated definitions of dynamic risk measures. One can even go further to define a reasonable class of dynamic risk measures for a model with incomplete information. Thorough investigations on Markov decision processes can be found e. g. in Hinderer (1970), Puterman (1994) or Hernández-Lerma and Lasserre (1996). Further applications beyond economics are possible in biology or telecommunications, for example.

An MDP is defined as follows. First, one needs a dynamical system of interest which can be influenced over time by a risk manager through the choice of certain decision variables. Then, one has to define the state space of the process of interest, the action space and the restriction set of admissible decision variables. Furthermore, a transition function and a transition law are introduced. They describe the realization of the (random) state of the process in the next period in case the former state and the current decision variable are given. Usually, it is assumed that in each period a certain reward depending on the current state and action is generated and that at the end of the finite time scale, there also is a terminal reward depending on the final state. The objective is now to find a policy, i. e. a sequence of decisions, which maximizes the expected accumulated reward. The main result of the theory is that this problem can be solved by the so-called *value iteration*, which is a sequence of one-period optimization problems. By this method, one can obtain the optimal target value and an optimal policy. Now we give an outline of this work and briefly describe how Markov decision theory can be applied in dynamic risk management. We will regularly state economic interpretations of important results. In particular, this concerns the structure of optimal investment policies derived in Chapter 2 and inequalities obtained when comparing values of dynamic risk measures.

In *Chapter 1* the general notion of static and conditional risk measures is introduced. We review the related literature and describe important examples such as Value-at-Risk and Average Value-at-Risk. Properties of risk measures are discussed and motivated economically. Furthermore, representation results from the literature are stated for coherent risk measures and for risk measures of Kusuoka-type. In particular, we stress the importance of Average Value-at-Risk, which will also become obvious in the consecutive chapters. Finally, we prove some simple measurability results for the conditional versions of Value-at-Risk and Average Value-at-Risk. This is essential when applying these risk measures in the construction of our dynamic risk measure in Chapter 4.

Chapter 2 is devoted to a natural application of Average Value-at-Risk, namely in the context of dynamic portfolio optimization. We consider a standard multi-period Cox-Ross-Rubinstein model with one bond and one risky asset, where an investor is able to invest his wealth into one risk-free bond and one risky asset. The following mean-risk optimization problems are investigated. First, we aim to minimize the risk represented by the Average Value-at-Risk of the final wealth under the constraint that at least a minimal expected final wealth is attained. This problem can be solved by a standard convex optimization approach via the Lagrange function and by formulating a Markov decision model. The convex optimization approach is possible due to convexity of Average Value-at-Risk. As a tool, we derive a generalization of a theorem by Runggaldier et al. (2002). In the same manner, the converse problem is solved, i. e. we maximize the expected final wealth under an Average Value-at-Risk constraint. We also give optimal investment strategies, showing that these are equal for both problems and of the structure of hedging strategies in the standard Cox-Ross-Rubinstein model. In the last part of the chapter, we formulate a variation of the latter problem by considering intermediate risk constraints when maximizing the expected final wealth. The idea is to ensure that – based on the current wealth – the risk of the relative gain or loss until the next period is bounded by a given risk level. Again, a Markov decision model yields the solution of the optimization problem. Instead of considering a Lagrangian approach, we have to adapt the restriction set of the related MDP and directly obtain the optimal policy, which is of a different structure than the one from above. Furthermore, the optimal target value has a different structure, too.

While so far only static and conditional risk measures have been considered, *Chapter 3* introduces the notion of *dynamic risk measures*. The demand for such tools is motivated by the following reasons based on arguments from Wang (1999). First, companies often face financial positions that have to be described by processes rather than by final payments. On the other hand, risk is not only to be measured at an initial date but should be adapted over time using the information modelled by a filtration which becomes available over time. First, we give a thorough overview on the related literature. Similar to Chapter 1 we describe important and economically motivated properties and state representation results for dynamic risk measures from the literature. It turns out that it is important to discern between law invariant measures and ones without this property, where the latter one seems to be the more natural one. Moreover, the notion of stability of sets of probability measures is closely related to the given representations. Concluding the chapter we show that this definition is equivalent to consistency of such sets defined in Riedel (2004).

The overview on the literature has shown that there is still a lack of reasonable and practicable dynamic risk measures. Hence, in *Chapter 4* a risk measure for processes by Pflug and Ruszczyński (2001) is generalized by constructing a dynamic version of this measure. First, we show that under certain conditions, the risk measure fulfills most of the properties introduced in the previous chapter. Since the dynamic risk measure is defined as an optimization problem, we aim to solve this and give a closed formula for the risk measure. We are able to do so in a Markovian environment by defining a related Markov decision model and applying the value iteration. In a further step, we consider the relationship of the dynamic risk measure with the stable representation results stated in Chapter 3. It turns out that this is only possible for income processes which are final values, but not for general income processes. However, one can derive a different stable representation for our dynamic risk measure. Concluding the chapter we prove some simple martingale properties which allow for a standard interpretation of the monetary values that the components of the risk measure represent.

The solution via Markov decision processes in Chapter 4 gives rise to the idea of generalizing the model to one with incomplete information in *Chapter 5*. More precisely, we assume that the generating variables of the model depend on a random parameter whose distribution is unknown. Now, the optimization problem in the definition of the dynamic risk measure from the previous chapter is transformed by introducing a so-called Bayesian optimization problem. The solution of the problem is obtained by extending the state space of the Markov decision model by a distributional component, representing the current estimation of the unknown distribution. It is well known that this model is equivalent to the Bayesian formulation. We use this approach to define a class of dynamic risk measures for such models in an analogous way as in Chapter 4. It can be observed that the models are identical if the parameter is known. Furthermore, we are able to solve the optimization in a binomial model in the case the unknown distribution is in the class of Beta distributions, compare also Attachment B. As a central result of this chapter we proof a comparison result for the two different risk measures in the binomial model. The necessary assumptions are fulfilled in the Cox–Ross–Rubinstein model, but not for a three period game by Artzner. However, due to the elementary character of the latter example, we can also give corresponding comparison results for the latter example.

Some basic notation is used throughout this work.

$\wp(E)$	the power set of an arbitrary set E
$\mathbb{N}$ ($\mathbb{N}_0$)	positive (non–negative) integers
$\mathbb{R}$ ($\mathbb{R}_+$, $\mathbb{R}_+^*$)	(non–negative, positive) real numbers
x^+ (x^-)	positive (negative) part of $x \in \mathbb{R}$, i. e. $x^+ = \max\{x, 0\}$ ($x^- = (-x)^+$)
$\mathcal{B}$	Borel sets on the real line
$\mathcal{B}_A$	Borel sets on a subset $A \subset \mathbb{R}$
$\overline{\mathcal{B}}$	Borel sets on $\overline{\mathbb{R}} = \mathbb{R} \cup \{\infty, -\infty\}$
$(\Omega, \mathcal{F}, \mathbb{P})$	basic probability space
$\mathbb{E}$	expectation with respect to $\mathbb{P}$
$\mathbb{E}_Q$	expectation with respect to another probability measure Q on $(\Omega, \mathcal{F})$
$\mathcal{Q}$	a set of probability measures on $(\Omega, \mathcal{F})$
$\mathcal{Q}^*$ ($\mathcal{Q}^e$)	a set of probability measures on $(\Omega, \mathcal{F})$ which are absolutely continuous with respect (equivalent) to $\mathbb{P}$
$\mathcal{P}(\Theta)$	the set of all probability measures on $\Theta \subset \mathbb{R}$
$\mathcal{G}$	a sub–σ–algebra of $\mathcal{F}$
$L^0(\Omega, \mathcal{G}, \mathbb{P})$	equivalence classes of $(\mathcal{G}, \mathcal{B})$–measurable risks.
$L^p(\Omega, \mathcal{G}, \mathbb{P})$	equivalence classes of $(\mathcal{G}, \mathcal{B})$–measurable and p–integrable (or bounded) risks, $p \in [1, \infty]$.
$\overline{L^p}(\Omega, \mathcal{G}, \mathbb{P})$	analogous equivalence classes of $(\mathcal{G}, \overline{\mathcal{B}})$–measurable risks
$X_n \xrightarrow{L^1} X$	convergence in L^1, i. e. $\lim_{n\to\infty} \mathbb{E}\,\lvert X_n - X\rvert = 0$
ess. sup, ess. inf	essential supremum (infimum) of a random variable or of a set of random variables, compare Attachment A
$\mathbb{1}_A(x) \in \{0, 1\}$	indicator function of a set A, i. e. $\mathbb{1}_A(x) = 1$ if and only if $x \in A$
$\mathcal{L}(X)$	the law of a random variable X
$\mathcal{L}(X \mid Y = y)$	the conditional law of a random variable given $\{Y = y\}$
$\square$	marks the end of a proof

1 Static and conditional risk measures

The mathematical notion risk in its most general definition corresponds to the uncertain value of a financial position at a future point of time denoted by T. This could be e. g. the price of an asset or an insurance claim, which a bank or an insurance company faces. In our setting, a positive value of the position is interpreted as an income whereas a negative value is a loss. Sometimes, the literature deals with the reverse definition. However, we follow the setting of the central works Artzner et al. (1999) and Acerbi and Tasche (2002). We assume that all occurring random variables already represent discounted values. In the first section of this chapter we introduce the underlying model and the notion of coherence. In the following two sections, relevant examples such as the Value-at-Risk and the Average Value-at-Risk and further properties are described. The last section deals with conditional (static) risk measures. These will be an essential tool for the construction of an explicit dynamic risk measure which will be introduced in Chapter 4.

A more practicably orientated version of the first three sections, which are essentially based on works as Artzner et al. (1999), Tasche (2002), Acerbi and Tasche (2002) and Szegö (2005), can be found in Bäuerle and Mundt (2005). Furthermore, several relevant applications are described there.

1.1 Model and definition

We now want to *measure the risk*. To this extend, let us formally define what a risk and a risk measure is. Assume that we are given some probability space $(\Omega, \mathcal{F}, \mathbb{P})$. All occurring equalities and inequalities among random variables are understood in the almost sure sense with respect to $\mathbb{P}$. We restrict ourselves to integrable risks since the Average Value-at-Risk, which we will use frequently throughout this work, can only be defined for those risks.

Definition 1.1. *A* risk *is a* $(\mathcal{F}, \mathcal{B})$*-measurable random variable* $X : \Omega \to \mathbb{R}$. *Denote with* $L^1(\Omega, \mathcal{F}, \mathbb{P})$ *the equivalence classes of integrable risks. A mapping* $\rho : L^1(\Omega, \mathcal{F}, \mathbb{P}) \to \overline{\mathbb{R}}$ *is called a* risk measure *if it is* monotone *and* translation invariant, *i. e.*

(MON) For all $X_1, X_2 \in L^1(\Omega, \mathcal{F}, \mathbb{P})$ it holds

$$X_1 \le X_2 \quad \Rightarrow \quad \rho(X_1) \ge \rho(X_2).$$

(TIN) For all $X \in L^1(\Omega, \mathcal{F}, \mathbb{P})$ and $c \in \mathbb{R}$, it holds

$$\rho(X + c) = \rho(X) - c.$$

The following class of convex risk measures is thoroughly investigated for bounded random variables in Föllmer and Schied (2004).

Definition 1.2. *A risk measure ρ is called* convex, *if we have the following:*

(CVX) For all $X_1, X_2 \in L^1(\Omega, \mathcal{F}, \mathbb{P})$ and $\lambda \in [0, 1]$ holds

$$\rho(\lambda X_1 + (1 - \lambda)X_2) \le \lambda\rho(X_1) + (1 - \lambda)\rho(X_2).$$

The property of convexity is a generalization of coherence, a notion that was discussed in the literature earlier than convexity. The leading work was done in Artzner et al. (1999), where risk measures on finite probability spaces where investigated.

Definition 1.3. *A convex risk measure ρ is called* coherent, *if it is* homogeneous*:*

(HOM) For all $X \in L^1(\Omega, \mathcal{F}, \mathbb{P})$ and $\lambda > 0$ it holds

$$\rho(\lambda X) = \lambda\rho(X).$$

Remark. If (HOM) is fulfilled, then convexity is equivalent to *subadditivity*, i. e.

(SUB) For all $X_1, X_2 \in L^1(\Omega, \mathcal{F}, \mathbb{P})$ it holds

$$\rho(X_1 + X_2) \le \rho(X_1) + \rho(X_2).$$

Consequently, coherence of a mapping $\rho : L^1(\Omega, \mathcal{F}, \mathbb{P}) \to \overline{\mathbb{R}}$ is usually defined by the four properties (MON), (TIN), (HOM) and (SUB). Note that a homogeneous mapping $\rho : L^1(\Omega, \mathcal{F}, \mathbb{P}) \to \mathbb{R}$ is always normalized, i. e. $\rho(0) = 0$. This can be seen by

$$\rho(0) = \rho(2 \cdot 0) = 2\rho(0).$$

Moreover, if ρ is translation invariant and normalized, we obtain for $c \in \mathbb{R}$

$$\rho(c) = \rho(0 + c) = \rho(0) - c = -c.$$

For $|\Omega| < \infty$, the main result in Artzner et al. (1999) was a representation theorem for coherent risk measures which showed that coherence does not determine a unique risk measure but rather a whole class. To derive a general representation result on $L^1(\Omega, \mathcal{F}, \mathbb{P})$, a technical continuity property has to be introduced. We now state the general version of the representation theorem which is from Inoue (2003).

Theorem 1.1. *A mapping* $\rho : L^1(\Omega, \mathcal{F}, \mathbb{P}) \to \overline{\mathbb{R}}$ *is a coherent risk measure and fulfills the* Fatou property

(FAT) For all $X, X_1, X_2, \dots \in L^1(\Omega, \mathcal{F}, \mathbb{P})$ such that $X_n \xrightarrow{L^1} X$, $n \to \infty$, it holds
$$\rho(X) \leq \liminf_{n\to\infty} \rho(X_n).$$

if and only if there exists a convex $L^1(\Omega, \mathcal{F}, \mathbb{P})$*-closed and* $L^\infty(\Omega, \mathcal{F}, \mathbb{P})$*-bounded set* $\mathcal{Q}$ *of probability measures on* $(\Omega, \mathcal{F})$ *which are absolutely continuous with respect to* $\mathbb{P}$ *such that*
$$\rho(X) = \sup_{Q\in\mathcal{Q}} \mathbb{E}_Q[-X], \quad X \in L^1(\Omega, \mathcal{F}, \mathbb{P}). \tag{1.1}$$

Remark. Before the general result was derived in Inoue (2003), the representation was proved for bounded random variables on general probability spaces in Delbaen (2002). The Fatou property then has to be formulated for uniformly bounded sequences that converge in probability. Generalizations of the result for convex risk measures can be found e. g. in Föllmer and Schied (2004).

Economic interpretation. An intensive discussion of the introduced properties can be observed in the literature. We give here a short summary of the main arguments. Fix a mapping $\rho : L^1(\Omega, \mathcal{F}, \mathbb{P}) \to \overline{\mathbb{R}}$ and, whenever necessary, a risk $X \in L^1(\Omega, \mathcal{F}, \mathbb{P})$.

The properties of monotonicity and translation invariance are well accepted. We consequently used them as central properties that a mapping should have if it is called a risk measure. First, obviously a risk that is pointwise not smaller than another risk should be assigned a value of the risk measure that is not larger. Hence, monotonicity is a natural assumption when measuring risk. Furthermore, translation invariance implies

$$\rho(X + \rho(X)) = \rho(X) - \rho(X) = 0, \tag{1.2}$$

hence providing an interpretation of $\rho(X)$ as a monetary value as follows. If we have $\rho(X) > 0$, this number can be seen as an amount of money that is invested in a risk-free bond (recall that we neglect the interest rate here). If both X and $\rho(X)$ are held in an aggregated portfolio, its value of the risk measure then becomes 0 by (1.2). On the other hand, if $\rho(X) \leq 0$, the amount $-\rho(X)$ can be subtracted from X with the overall risk still being 0.

The discussion about homogeneity is more ambiguous. This property seems to make sense if $\rho(X)$ is indeed seen as a monetary value. By homogeneity, the value of the risk measure changes proportionally if the financial position X is calculated in a different currency. An argument against this property is mentioned in Föllmer and Schied (2004). They claim that there are many situations where the value of the risk measure does not increase linearly if this is the case for the position $-X$.

There is no more extensive study of this argument. We conclude that assuming homogeneity or not is the essential difference of the two worlds of coherent and convex risk measures respectively.

The properties of subadditivity and, in a similar way of convexity, are again quite well–established. There is some discussion though because the famous Value–at–Risk is not subadditive whereas this risk measure is broadly accepted in the banking industry. This is further strengthened by the proclamation of Value–at–Risk within the context of Basel II. The standard argument in favor of subadditivity is that it encourages diversification. More precisely, if a risk measure ρ is not subadditive, it could happen that a company which uses ρ can be led to split the company in order to reduce the total capital requirement. However, this is not desired from a regulatory point of view. Indeed, the Swiss "Federal Office of Private Insurers" is on its way to regulate by law the coherent risk measure Average Value–at–Risk for determining the target value of an insurance company. This is also considered in the negotiations about Solvency 2, while the banking sector still favors Value–at–Risk.

Additionally, subadditivity is a favorable property for the company itself and its policy holders. First, it is useful to obtain a bound for the risk of the whole company by summing up the risk of the different branches. Moreover, it implies that the risk of a whole group of policy holders is smaller than the sum of the singular risks.

We have seen which central properties of risk measures can be motivated by mathematical and economical aspects. Now, we are going to introduce the aforementioned risk measures Value–at–Risk and Average Value–at–Risk and discuss its properties. Furthermore, some more classes of risk measures, namely law invariant and comonotone risk mappings will be investigated in Section 1.3.

1.2 Examples

Let us now define some examples of risk measures. In particular, we formally define Value–at–Risk (V@R) and Average Value–at–Risk (AV@R).

The worst–case–measure and the negative expectation

Two simple coherent, but not very practicably relevant risk measures can be defined as follows:

$$\begin{aligned} \rho_{\mathrm{WC}}(X) &:= -\operatorname{ess.inf}(X) = \operatorname{ess.sup}(-X), \\ \rho_{\mathrm{EV}}(X) &:= \mathbb{E}\,[-X], \end{aligned}$$

where $X \in L^1(\Omega, \mathcal{F}, \mathbb{P})$. The worst–case–measure ρ_{WC} becomes ∞ for unbounded random variables, such that in this case, a risk measure ρ should be allowed to attain values in $\overline{\mathbb{R}}$. It is too conservative to be used in a reasonable way. The same conclusion holds for the negative expectation ρ_{EV} because it does not contain any relevant information about the riskiness of the underlying financial position X.

Both risk measures fit in the context of Theorem 1.1 with

$$\mathcal{Q}_{\mathrm{WC}} := \mathcal{Q}^* \quad \text{and} \quad \mathcal{Q}_{\mathrm{EV}} := \{\mathbb{P}\},$$

where $\mathcal{Q}^*$ is the set of all probability measures that are absolutely continuous with respect to $\mathbb{P}$.

Value–at–Risk

A classic and famous – though strongly criticized in works as Artzner et al. (1999), Acerbi and Nordio (2001) or Acerbi and Tasche (2002) due to its lack of subadditivity – homogeneous risk measure is Value–at–Risk. Defining upper and lower quantiles of the distribution function $F_Z = \mathbb{P}(Z \leq \cdot)$ of a random variable $Z : \Omega \to \mathbb{R}$ via

$$q_\gamma^+(Z) = \inf\{z \in \mathbb{R} \mid F_Z(z) > \gamma\} = \sup\{z \in \mathbb{R} \mid F_Z(z) \leq \gamma\}, \quad \gamma \in (0,1),$$

and

$$q_\gamma^-(Z) = \inf\{z \in \mathbb{R} \mid F_Z(z) \geq \gamma\} = \sup\{z \in \mathbb{R} \mid F_Z(z) < \gamma\}, \quad \gamma \in (0,1),$$

the Value–at–Risk for a level $\gamma \in (0,1)$ of a risk X is the lower γ–quantile of $-X$, formally

$$\mathrm{V@R}_\gamma(X) := -q_{1-\gamma}^+(X) = q_\gamma^-(-X) = \inf\{x \in \mathbb{R} \mid \mathbb{P}(-X \leq x) \geq \gamma\}, \gamma \in (0,1).$$

Consequently, $\mathrm{V@R}_\gamma(X)$ represents the smallest monetary value such that $-X$ does not exceed this value at least with probability γ. Hence, we are interested in calculating the Value–at–Risk for large values of γ, e. g. $\gamma = 0.95$ or $\gamma = 0.99$. There are two main disadvantages of Value–at–Risk. First, we have already mentioned that it is in general not subadditive, see Acerbi and Tasche (2002). However, in the family of elliptical distributions, Value–at–Risk is subadditive such that it is not always an inappropriate risk measure. Another disadvantage is the fact that Value–at–Risk ensures that our future loss stays below a certain level with some high probability γ, whereas it does not take the amount of the loss into account – if it occurs. More generally, the distribution of X above this level is not relevant. This might not be interesting for stake holders, but is important for regulatory institutions for example. To overcome these drawbacks, the Average Value–at–Risk was developed. It is worth mentioning that another drawback is that Value–at–Risk is not sensitive to small changes of the safety level γ, i. e. is not continuous with respect to this probability.

Average Value–at–Risk

This risk measure can be defined by using Value–at–Risk. Indeed, for $\gamma \in (0,1)$ it represents an average of the Value–at–Risk to all safety levels larger than γ, formally

$$\mathrm{AV@R}_\gamma(X) := \frac{1}{1-\gamma}\int_\gamma^1 \mathrm{V@R}_u(X)\mathrm{d}u, \quad \gamma \in (0,1), \tag{1.3}$$

for $X \in L^1(\Omega, \mathcal{F}, \mathbb{P})$. The definition, which is continuous and strictly increasing in the safety level γ, is also valid for $\gamma = 0$ and we obtain $\mathrm{AV@R}_0(X) = \rho_{\mathrm{EV}}(X)$. Furthermore, we set $\mathrm{AV@R}_1(X) = \rho_{\mathrm{WC}}(X)$, compare Kusuoka (2001). Note that we always have

$$\mathrm{AV@R}_\gamma(X) \geq \mathrm{V@R}_\gamma(X), \quad \gamma \in [0,1].$$

However, Average Value–at–Risk can attain the value ∞ whereas Value–at–Risk is always finite.

Coherence of Average Value–at–Risk is ensured by the following result which shows that the risk measure fits into the setting of Theorem 1.1.

Theorem 1.2. *Let $\gamma \in [0,1]$. Then $\mathrm{AV@R}_\gamma$ is a coherent risk measure with the Fatou property. It holds*

$$\mathrm{AV@R}_\gamma(X) = \sup_{Q \in \mathcal{Q}_\gamma} \mathbb{E}_Q[-X], \quad X \in L^1(\Omega, \mathcal{F}, \mathbb{P}),$$

with the set of probability measures

$$\mathcal{Q}_\gamma := \left\{ Q \ll \mathbb{P} \mid \frac{\mathrm{d}Q}{\mathrm{d}\mathbb{P}} \leq \frac{1}{1-\gamma} \right\}, \quad \gamma \in [0,1]. \tag{1.4}$$

Proof. See Theorem 4.47 in Föllmer and Schied (2004). □

By definition, it can be observed that Average Value–at–Risk indeed takes into consideration how high a potential loss above the level $\mathrm{V@R}_\gamma(X)$ can be. Besides the name Average Value–at–Risk, also notions as Conditional Value–at–Risk or Expected Shortfall can be found frequently in the literature. Usually, they denote the same risk measure. However, sometimes care has to be taken how the authors define their risk measures because the same name could be used for different risk measures. If X has an absolutely continuous distribution, it holds

$$\mathrm{AV@R}_\gamma(X) = \mathbb{E}\left[-X \mid -X \geq \mathrm{V@R}_\gamma(X)\right],$$

which is sometimes claimed to be a coherent risk measure and called Average Value–at–Risk, which is wrong though for general distributions. We restrict ourselves to the name Average Value–at–Risk for the risk measure defined above because the name intuitively fits with the definition (1.3). Moreover, the name "Conditional Value–at–Risk" collides with the notion of conditional risk measures treated in Section 1.4.

The definition (1.3) still uses Value-at-Risk. For certain distributions, this might be numerically hard to obtain. But by Uryasev and Rockafellar (2002), Average Value-at-Risk can be represented by a simple convex optimization problem. Fix $\gamma \in (0,1)$ and $X \in L^1(\Omega, \mathcal{F}, \mathbb{P})$. Then we have

$$\mathrm{AV@R}_\gamma(X) = \inf_{b\in\mathbb{R}} \left\{ b + \frac{1}{1-\gamma} \mathbb{E}\left[(-X-b)^+\right] \right\}. \tag{1.5}$$

In fact, the infimum is actually attained in $b^* = \mathrm{V@R}_\alpha(X)$. Hence

$$\mathrm{AV@R}_\gamma(X) = \mathrm{V@R}_\gamma(X) + \frac{1}{1-\gamma} \mathbb{E}\left[(-X - \mathrm{V@R}_\gamma(X))^+\right]. \tag{1.6}$$

We have noted above that $\mathrm{AV@R}_\gamma(X)$ and $\mathbb{E}[-X \mid -X \geq \mathrm{V@R}_\gamma(X)]$ do not coincide. More precisely, it holds by Corollary 4.3 in Acerbi and Tasche (2002)

$$\begin{aligned} &\mathrm{AV@R}_\gamma(X) \\ &= \frac{1}{1-\gamma}\Big(\mathbb{E}\left[-X \cdot \mathbb{1}_{\{-X \geq \mathrm{V@R}_\gamma(X)\}}\right] - \mathrm{V@R}_\gamma(X)\big(\gamma - \mathbb{P}(-X < \mathrm{V@R}_\gamma(X))\big)\Big). \end{aligned}$$

The relation of Value-at-Risk and Average Value-at-Risk is emphasized by Figure 1.1.

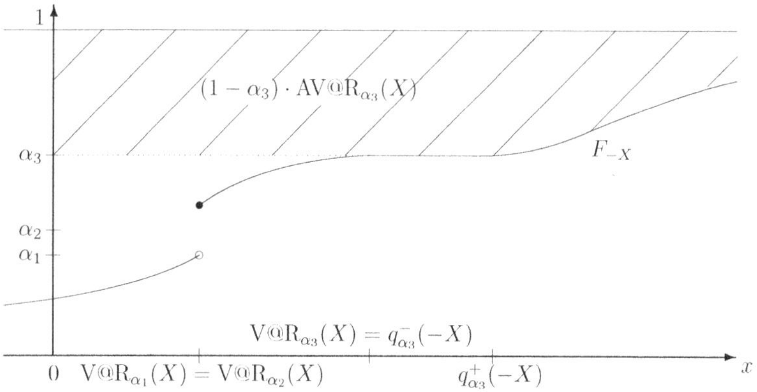

Figure 1.1: V@R and AV@R

We conclude that Average Value-at-Risk is a coherent risk measure which is more conservative than Value-at-Risk, two important facts that make it a reasonable tool for measuring risk. We have emphasized above that a certain conditional expectation can be mistaken for Average Value-at-Risk. Hence, let us take a closer look at some related risk measures to conclude this section.

Tail Conditional Expectation or Expected Shortfall

We can define the so-called Tail Conditional Expectation (TCE) for upper and lower quantiles and the Expected Shortfall (ES). Let $\gamma \in (0,1)$ and set

$$\begin{aligned} \mathrm{TCE}_\gamma^-(X) &:= \mathbb{E}[-X \mid X \leq q_{1-\gamma}^-(X)] = \mathbb{E}[-X \mid -X \geq q_\gamma^+(-X)], \\ \mathrm{TCE}_\gamma^+(X) &:= \mathrm{ES}_\gamma(X) := \mathbb{E}[-X \mid X \leq q_{1-\gamma}^+(X)] = \mathbb{E}[-X \mid -X \geq q_\gamma^-(-X)] \end{aligned}$$

for $X \in L^1(\Omega, \mathcal{F}, \mathbb{P})$. By definition, we have

$$\mathrm{TCE}_\gamma^+(X) \leq \mathrm{TCE}_\gamma^-(X)$$

and

$$\mathrm{TCE}_\gamma^+(X) = \mathbb{E}[-X \mid -X \geq \mathrm{V@R}_\gamma(X)].$$

Hence, some older works claimed $\mathrm{TCE}_\gamma^+(X) = \mathrm{AV@R}_\gamma(X)$. However it was shown in Acerbi and Tasche (2002) that in general only

$$\mathrm{TCE}_\gamma^+(X) \leq \mathrm{TCE}_\gamma^-(X) \leq \mathrm{AV@R}_\gamma(X) \tag{1.7}$$

holds and that the inequalities can be strict even for non-academic distributions. On the other hand, we have equality of all terms in (1.7) if X has a continuous distribution. The cited work gives a thorough overview on the relationship of all the risk measures introduced above.

1.3 Further properties and representations

In Kusuoka (2001), some more interesting properties are defined which are fulfilled by Value-at-Risk and Average Value-at-Risk. Two risks $X_1, X_2 \in L^1(\Omega, \mathcal{F}, \mathbb{P})$ are called *comonotone* if there exist non-decreasing $(\mathcal{B}_{[0,1]}, \mathcal{B})$-measurable functions $f_1, f_2 : [0,1] \to \mathbb{R}$ and a random variable U on $(\Omega, \mathcal{F}, \mathbb{P})$ with $\mathcal{L}(U) = \mathcal{U}(0,1)$ such that

$$X_1 = f_1(U), \quad X_2 = f_2(U).$$

Definition 1.4. *Let ρ be a risk measure.*

(NEG) ρ is called *negative* if for all risks $X \in L^1(\Omega, \mathcal{F}, \mathbb{P})$ we have

$$X \geq 0 \quad \Rightarrow \quad \rho(X) \leq 0.$$

(LIN) ρ is called *law invariant* if for all $X_1, X_2 \in L^1(\Omega, \mathcal{F}, \mathbb{P})$ it holds

$$\mathcal{L}(X_1) = \mathcal{L}(X_2) \quad \Rightarrow \quad \rho(X_1) = \rho(X_2).$$

(COM) ρ is called *comonotone additive* if for all risks $X_1, X_2 \in L^1(\Omega, \mathcal{F}, \mathbb{P})$ we have

$$X_1, X_2 \text{ comonotone} \quad \Rightarrow \quad \rho(X_1 + X_2) = \rho(X_1) + \rho(X_2).$$

Note that by a simple and standard argument, negativity and subadditivity imply monotonicity:

Let ρ be a negative and subadditive risk measure and let $X_1, X_2 \in L^1(\Omega, \mathcal{F}, \mathbb{P})$ with $X_1 \leq X_2$. Then we obtain first by subadditivity and then by negativity

$$\rho(X_2) = \rho(X_2 - X_1 + X_1) \leq \rho(X_2 - X_1) + \rho(X_1) \leq 0 + \rho(X_1) = \rho(X_1).$$

Hence, ρ is monotone.

The other two properties now lead to interesting subclasses of the class of coherent risk measures. The law invariant ones are characterized simply by using the Average Value at Risk and a subset of the class $\mathcal{P}(0, 1)$ of all probability measures on $(0, 1)$. If the subset contains only one probability measures on $(0, 1)$ we obtain the class of law invariant and comonotone risk measures. The interval is interpreted as the set of the safety levels γ for which the Average Value at Risk is defined. The following representation results were first proven by Kusuoka (2001) for bounded random variables and then generalized by Bäuerle and Müller (2006) for integrable ones.

Theorem 1.3. *A mapping $\rho : L^1(\Omega, \mathcal{F}, \mathbb{P}) \to \overline{\mathbb{R}}$ is a law invariant coherent risk measure with the Fatou property if and only if there exists a compact and convex set $\mathcal{M} \subset \mathcal{P}(0, 1)$ such that*

$$\rho(X) = \sup_{\mu \in \mathcal{M}} \int_0^1 \mathrm{AV@R}_\gamma(X) \mu(\mathrm{d}\gamma), \quad X \in L^1(\Omega, \mathcal{F}, \mathbb{P}).$$

We say that a risk measure with the above properties is of Kusuoka type.

Remark. The result can be generalized for convex risk measures, compare Theorem 4.54 in Föllmer and Schied (2004).

Adding the property of comonotone additivity yields the following representation.

Theorem 1.4. *A mapping $\rho : L^1(\Omega, \mathcal{F}, \mathbb{P}) \to \overline{\mathbb{R}}$ is a comonotone additive risk measure of Kusuoka type if and only if there exists $\mu \in \mathcal{P}(0, 1)$ such that*

$$\rho(X) = \int_0^1 \mathrm{AV@R}_\gamma(X) \mu(\mathrm{d}\gamma), \quad X \in L^1(\Omega, \mathcal{F}, \mathbb{P}).$$

By Theorem 9 in Kusuoka (2001), we have

$$\mathrm{AV@R}_\gamma(X) = \inf\{\rho(X) \mid \rho \text{ of Kusuoka–type and } \rho(X) \geq \mathrm{V@R}_\gamma(X)\},$$

for every $X \in L^1(\Omega, \mathcal{F}, \mathbb{P})$ and $\gamma \in (0,1)$. The proof, which is given there only in the case $X \in L^\infty(\Omega, \mathcal{F}, \mathbb{P})$, can be adapted exactly for integrable risks, based on Theorem 1.3. Hence, Average Value–at–Risk is the smallest coherent risk measure dominating Value–at–Risk.

In Jouini et al. (2006) it was further shown for bounded random variables that every law invariant convex risk measure has the Fatou–property. Note that in their notation, $-\rho$ is a monetary utility function. As far as we know the generalization to unbounded but integrable risks has not been proved yet. But this is not supposed to be part of this work.

Example 1.1. As a simple example for a law invariant and comonotone coherent risk measure we can consider for $N \in \mathbb{N}$ a sequence $(\alpha_j)_{j=1\dots,N} \subset [0,1)$ of safety levels and a sequence of weight factors $(p_j)_{j=1,\dots,N} \subset (0,1)$ such that $\sum_{j=1}^N p_j = 1$ and define

$$\rho(X) := \sum_{j=1}^N p_j \cdot \mathrm{AV@R}_{\alpha_j}(X), \quad X \in L^1(\Omega, \mathcal{F}, \mathbb{P}).$$

Accepting the two central properties law invariance and comonotone additivity from above as reasonable provides another argument for the importance of Average Value–at–Risk, because it generates the resulting subclasses of coherent risk measures. Hence, let us shortly investigate the economic background of the properties from Definition 1.4.

Economic interpretation. Negativity (which is implied by monotonicity if the risk measure is normalized) is a natural property. If a risk is non–negative no loss can occur, hence the risk measure should not attain a positive value for this risk.

An argument in favor of law invariance (at least in the static setting) can be found in Wang et al. (1997). If two risks follow the same law, they should be assigned the same value of the risk. This is because this value should depend on the probability distribution of the risk X rather than on the actual state of the world that leads to a realization of the financial position X.

Finally, comonotone additivity reflects a property that seems natural when rating a portfolio of two comonotone risks X_1 and X_2 by a coherent risk measure ρ. More precisely, if X_1 and X_2 are held together in one portfolio, the company faces a more risky situation as if these "coupled" risks were held separately. Hence, we should have

$$X_1, X_2 \text{ comonotone} \quad \Rightarrow \quad \rho(X_1) + \rho(X_2) \leq \rho(X_1 + X_2).$$

Together with the reverse inequality which is fulfilled by subadditivity, this leads to the property of comonotone additivity.

The conclusion of Section 1.2 and this section is the importance of Average Value at Risk. First, it has a economic interpretation and fulfills all relevant properties of a risk measure. Moreover, it can be used to generate further risk measures with interesting properties. In Chapter 2, Average Value at Risk will be implemented in classic portfolio optimization problems, either as target value or within the constraints. On the other hand, using Average Value at Risk we are able to define a dynamic risk measure for income processes and extend this to models with incomplete information. This will be done in Chapters 4 and 5. As a preparation, we need some remarks on conditional static risk measures in the following section.

1.4 Remarks on conditional risk measures and measurability

In this section, we briefly want to examine the notion of a conditional risk measure. The necessity for this comes from situations when at time $t = 0$ some information about the underlying risk is already available. More formally, we can interpret the static model from the previous sections in the sense that at time $t = 0$, we have no information represented by the trivial σ–algebra $\mathcal{F}_0 := \{\emptyset, \Omega\}$ and at time $t = T$, we have the information represented by the σ–algebra $\mathcal{F}$. An underlying risk X then is a $(\mathcal{F}, \mathcal{B})$–measurable random variable whereas the corresponding value of a risk measure $\rho(X)$ is, being a constant, $(\mathcal{F}_0, \mathcal{B})$–measurable. Now, we can model more available information at time $t = 0$ by replacing $\mathcal{F}_0$ with a general sub–σ–algebra $\mathcal{G} \subset \mathcal{F}$, which we fix throughout the rest of this section. Hence, the value of the risk measure should be allowed to be a $(\mathcal{G}, \mathcal{B})$–measurable random variable. This approach will be very important when considering dynamic risk measures in Chapter 3.

We will now describe the setting from works as Bion-Nadal (2004), Detlefsen and Scandolo (2005) and Ruszczyński and Shapiro (2006) and introduce conditional versions of the most important examples treated in Section 1.2 because these will frequently be used in the dynamic setting.

Definition 1.5. *A mapping* $\rho(\cdot \,|\, \mathcal{G}) : L^1(\Omega, \mathcal{F}, \mathbb{P}) \to \overline{L^0}(\Omega, \mathcal{G}, \mathbb{P})$ *is called a* conditional risk measure *if it is monotone and conditional translation invariant, i. e.*

(MON) For all $X_1, X_2 \in L^1(\Omega, \mathcal{F}, \mathbb{P})$ it holds

$$X_1 \leq X_2 \quad \Rightarrow \quad \rho(X_1 \,|\, \mathcal{G}) \geq \rho(X_2 \,|\, \mathcal{G}).$$

(CTI) For all $X \in L^1(\Omega, \mathcal{F}, \mathbb{P})$ and $Z \in L^1(\Omega, \mathcal{G}, \mathbb{P})$, it holds

$$\rho(X + Z \,|\, \mathcal{G}) = \rho(X \,|\, \mathcal{G}) - Z.$$

In addition, ρ is called coherent, *if it is conditional homogeneous and subadditive, i. e.*

(CHO) For all $X \in L^1(\Omega, \mathcal{F}, \mathbb{P})$ and $\Lambda \in L^1(\Omega, \mathcal{G}, \mathbb{P})$ with $\Lambda \geq 0$ it holds

$$\rho(\Lambda X \mid \mathcal{G}) = \Lambda \rho(X \mid \mathcal{G}).$$

(SUB) For all $X_1, X_2 \in L^1(\Omega, \mathcal{F}, \mathbb{P})$ it holds

$$\rho(X_1 + X_2 \mid \mathcal{G}) \leq \rho(X_1 \mid \mathcal{G}) + \rho(X_2 \mid \mathcal{G}).$$

Remark. Other properties such as convexity can also be defined analogously to the unconditional setting. We skip a more explicit investigation since we do not need the details here. In the aforementioned works, various representation theorems similar to Theorem 1.1 are derived. Again, it is shown that certain properties of a conditional risk measure $\rho(\,\cdot \mid \mathcal{G})$ are equivalent to $\rho(\,\cdot \mid \mathcal{G})$ being of the form

$$\rho(X \mid \mathcal{G}) = \operatorname{ess.\,sup}_{Q \in \mathcal{Q}} \mathbb{E}_Q[-X \mid \mathcal{G}] \tag{1.8}$$

for a set $\mathcal{Q}$ of probability measures being absolutely continuous with respect to $\mathbb{P}$.

The simplest example for a coherent conditional risk measure is the negative conditional expectation defined via

$$\rho_{\mathrm{EV}}(X \mid \mathcal{G}) := \mathbb{E}\,[-X \mid \mathcal{G}], \quad X \in L^1(\Omega, \mathcal{F}, \mathbb{P}).$$

Furthermore, we can introduce conditional versions of Value-at-Risk and Average Value-at-Risk which coincide with the usual versions if $\mathcal{G} = \{\emptyset, \Omega\}$. Note that since we consider only real valued random variables, a regular conditional distribution of X under $\mathcal{G}$ exists for $X \in L^1(\Omega, \mathcal{F}, \mathbb{P})$, which can be verified by Theorem 6.3 in Kallenberg (2002). More precisely, there exists a probability kernel $P^{(X,\mathcal{G})} : \Omega \times \mathcal{B} \to [0, 1]$ from $(\Omega, \mathcal{G})$ to $(\mathbb{R}, \mathcal{B})$ such that

$$P^{(X,\mathcal{G})}(\,\cdot\,, B) = \mathbb{P}(X \in B \mid \mathcal{G})(\,\cdot\,) \quad \mathbb{P}\text{-almost surely}$$

for every $B \in \mathcal{B}$. Now, using law invariance, the following is well defined for $\gamma \in (0, 1)$ and $\omega \in \Omega$:

$$\mathrm{V@R}_\gamma(X \mid \mathcal{G})(\omega) := \mathrm{V@R}_\gamma(P^{(X,\mathcal{G})}(\omega, \,\cdot\,)) := \inf\{x \in \mathbb{R} \mid \mathbb{P}(-X \leq x \mid \mathcal{G})(\omega) \geq \gamma\},$$

although it is not clear yet if $\mathrm{V@R}_\gamma(X \mid \mathcal{G})$ is indeed $(\mathcal{G}, \mathcal{B})$-measurable. We will verify this in a lemma below. Before doing so, let us introduce the corresponding conditional version of Average Value-at-Risk. This is inspired by (1.5) and can be found e. g. in Ruszczyński and Shapiro (2006). Again, let $X \in L^1(\Omega, \mathcal{F}, \mathbb{P})$, $\gamma \in (0, 1)$ and $\omega \in \Omega$. Set

$$\mathrm{AV@R}_\gamma(X \mid \mathcal{G})(\omega) := \inf_{Z\ \mathcal{G}\text{-meas.}} \mathbb{E}\left[Z + \frac{1}{1-\gamma}(-X - Z)^+ \mid \mathcal{G}\right](\omega). \tag{1.9}$$

Now, measurability of the defined conditional versions is provided by the following result.

Lemma 1.1. *Let $X \in L^1(\Omega, \mathcal{F}, \mathbb{P})$ and $\gamma \in (0,1)$. Then it holds:*

(i) $\mathrm{V@R}_\gamma(X \mid \mathcal{G})$ *is* $(\mathcal{G}, \mathcal{B})$*–measurable. Moreover,* $\mathrm{V@R}_\gamma(\cdot \mid \mathcal{G})$ *is a homogeneous conditional risk measure.*

(ii) $\mathrm{AV@R}_\gamma(X \mid \mathcal{G})$ *is* $(\mathcal{G}, \overline{\mathcal{B}})$*–measurable and we have*

$$\mathrm{AV@R}_\gamma(X \mid \mathcal{G}) = \mathrm{V@R}_\gamma(X \mid \mathcal{G}) + \frac{1}{1-\gamma}\mathbb{E}\left[(-X - \mathrm{V@R}_\gamma(X \mid \mathcal{G}))^+ \mid \mathcal{G}\right].$$

Moreover, $\mathrm{AV@R}_\gamma(\cdot \mid \mathcal{G})$ *is a coherent conditional risk measure.*

Proof. (i) We have to show $(\mathrm{V@R}_\gamma(X \mid \mathcal{G}))^{-1}(B) \in \mathcal{G}$ for every $B \in \overline{\mathcal{B}}$. It suffices to consider Borel sets of the type $B = (-\infty, x]$ for some $x \in \mathbb{R}$. We obtain

$$\begin{aligned}
\big(\mathrm{V@R}_\gamma(X \mid \mathcal{G})\big)^{-1}\big((-\infty, x]\big) &= \{\omega \in \Omega \mid \mathrm{V@R}_\gamma(X \mid \mathcal{G})(\omega) \le x\} \\
&= \{\omega \in \Omega \mid \mathbb{P}(X > -x \mid \mathcal{G})(\omega) \ge \gamma\} \\
&= \{\omega \in \Omega \mid \mathbb{P}(X \in (-\infty, -x] \mid \mathcal{G})(\omega) \le 1-\gamma\} \\
&= \big(\mathbb{P}(X \in (-\infty, -x] \mid \mathcal{G})\big)^{-1}\big([0, 1-\gamma]\big) \in \mathcal{G}.
\end{aligned}$$

The last step follows since, by definition, $\mathbb{P}(X \in B \mid \mathcal{G})$ is $(\mathcal{G}, \mathcal{B}_{[0,1]})$–measurable.

Monotonicity of $\mathrm{V@R}_\gamma(\cdot \mid \mathcal{G})$ can now be obtained directly from the definition as in the unconditional case. Conditional translation invariance and homogeneity can be proved in similar ways, so we restrict ourselves to translation invariance. Let $Z \in L^1(\Omega, \mathcal{G}, \mathbb{P})$. Proposition 2.13 in Yong and Zhou (1999) yields for almost every $\omega \in \Omega$:

$$\begin{aligned}
\mathrm{V@R}_\gamma(X + Z \mid \mathcal{G})(\omega) &= \inf\{x \in \mathbb{R} \mid \mathbb{P}(-X - Z \le x \mid \mathcal{G})(\omega) \ge \gamma\} \\
&= \inf\{x \in \mathbb{R} \mid \mathbb{P}(-X \le x + Z(\omega) \mid \mathcal{G})(\omega) \ge \gamma\} \\
&= \inf\{x - Z(\omega) \in \mathbb{R} \mid \mathbb{P}(-X \le x \mid \mathcal{G})(\omega) \ge \gamma\} \\
&= \mathrm{V@R}_\gamma(X \mid \mathcal{G})(\omega) - Z(\omega).
\end{aligned}$$

(ii) Measurability of $\mathrm{AV@R}_\gamma(X \mid \mathcal{G})$ was already shown in Ruszczyński and Shapiro (2006), where it was proved that the conditional Average Value–at–Risk can be represented in the form (1.8) with the same sets as in the unconditioned case introduced in (1.4). Furthermore, this already yields coherence of the conditional Average Value–at–Risk. However, measurability can alternatively be obtained if we show that the optimization problem in (1.9) is solved by $Z^* = \mathrm{V@R}_\gamma(X \mid \mathcal{G})$. By Proposition 2.13 in Yong and Zhou (1999) we have

$$\mathrm{AV@R}_\gamma(X \mid \mathcal{G})(\omega) = \inf_{Z\ \mathcal{G}\text{–meas.}} \left\{ Z(\omega) + \frac{1}{1-\gamma}\mathbb{E}\left[(-X - Z(\omega))^+ \mid \mathcal{G}\right](\omega) \right\}$$

for almost every $\omega \in \Omega$. Consequently, the target value depends for every $(\mathcal{G}, \mathcal{B})$-measurable Z only on $Z(\omega)$ and not on $Z(\omega')$, $\omega' \neq \omega$. Hence, by the solution in the static case in (1.6), the infimum is attained for every $(\mathcal{G}, \mathcal{B})$-measurable Z such that $Z(\omega) = \mathrm{V@R}_\gamma(X \mid \mathcal{G})(\omega)$. This is e. g. fulfilled for $Z = \mathrm{V@R}_\gamma(X \mid \mathcal{G})$, which is an admissible choice by part (i), therefore yielding the assertion. □

Since the unconditional versions of Value-at-Risk and Average Value-at-Risk represent law invariant risk measures, we can also interpret them as functions on the set of all probability distributions on $\mathbb{R}$ with finite mean rather than on $L^1(\Omega, \mathcal{F}, \mathbb{P})$. In this setting, the following observation is simple but important.

Corollary 1.1. *If it holds $\mathcal{G} = \sigma(Y)$ for some real-valued random variable Y on $(\Omega, \mathcal{F}, \mathbb{P})$, we obtain for almost every $\omega \in \Omega$*

$$\begin{aligned} \mathrm{V@R}_\gamma(X \mid Y)(\omega) &= \mathrm{V@R}_\gamma(\mathcal{L}(X \mid Y = Y(\omega))), \\ \mathrm{AV@R}_\gamma(X \mid Y)(\omega) &= \mathrm{AV@R}_\gamma(\mathcal{L}(X \mid Y = Y(\omega))). \end{aligned}$$

Proof. Let $\omega \in \Omega$. Then we have as a direct consequence of the factorization lemma for conditional expectations $P^{(X,\sigma(Y))}(\omega, \cdot) = \mathcal{L}(X \mid Y = Y(\omega))$. Hence, the assertion for Value-at-Risk is clear. But for Average Value-at-Risk, the same follows directly from the representation in Lemma 1.1 (ii) □

Remark. More intuitively, we write in the setting of the previous corollary for every law invariant risk measure ρ

$$\rho(X \mid Y = y) := \rho(\mathcal{L}(X \mid Y = y)), \quad y \in \mathrm{Range}(Y).$$

Concluding this section, we give another example of a conditional risk measure which is just a mixture of $\rho_{\mathrm{EV}}(\cdot \mid \mathcal{G})$ and $\mathrm{AV@R}_\gamma(\cdot \mid \mathcal{G})$. It will be a major tool in Chapters 4 and 5.

Example 1.2. Define for $X \in L^1(\Omega, \mathcal{F}, \mathbb{P})$, $\gamma \in (0,1)$ and $p \in (0,1)$

$$\rho(X \mid \mathcal{G}) := p \cdot \mathbb{E}\,[-X \mid \mathcal{G}] + (1-p) \cdot \mathrm{AV@R}_\gamma(X \mid \mathcal{G}).$$

We obtain in a similar calculation as done in the proof of Lemma 1 in Pflug and Ruszczyński (2001):

$$\rho(X \mid \mathcal{G}) = \inf_{Z\ \mathcal{G}\text{-meas.}} \mathbb{E}\left[Z + \frac{1-\gamma p}{1-\gamma}(-X-Z)^+ - p(-X-Z)^- \mid \mathcal{G}\right],$$

where the infimum is again, similar as in the proof of part (ii) of Lemma 1.1, attained in $Z = \mathrm{V@R}_\gamma(X \mid \mathcal{G})$.

If $\mathcal{G} = \mathcal{F}_0$, the static risk measure $\rho := \rho(\cdot \mid \mathcal{G})$ even is a comonotone risk measure of Kusuoka-type by Example 1.1.

2 Portfolio optimization with constraints in a binomial model

A classic problem in financial mathematics is mean–risk optimization, in particular when the wealth is modelled by a portfolio. For example, one aims to find an optimal investment strategy into one risk–free bond and one risky asset. In the standard theory, there are two important optimization problems with constraints. First, one could aim to minimize the risk of the portfolio under some constraint that ensures a certain minimum expected return. The other approach could be to maximize this expected return under some risk constraint. We will show how these problems are related in a binomial model and solve the corresponding optimization problems by introducing a classic Markov decision model. Furthermore, we will consider a model that uses intermediate risk constraints. The idea for the latter model is inspired by Cuoco et al. (2001) and Pirvu (2007), where continuous time models with Value–at–Risk–constraints are investigated.

2.1 The model

We consider a classic multi period Cox–Ross–Rubinstein model (CRRM) on a given probability space $(\Omega, \mathcal{F}, \mathbb{P})$ with one risk–free bond and one risky asset. Our main concern is the final value of the portfolio, i. e. we want to maximize its expected utility or minimize its risk. In the classic theory of mean–risk optimization, the risk is measured by the variance. But since it is not a very reasonable risk measure, we replace it with the Average Value–at–Risk. We have seen in the previous chapter that this is a economically founded choice.

The Markov decision model defined below is similar to the one implicitly used in Runggaldier et al. (2002) and Favero and Vargiolu (2006). The time horizon is finite and discrete. Hence, the trading takes place at times $t \in \{0, 1, \dots, T\}$ for some $T \in \mathbb{N}$. Assuming that the bond is constantly equal to 1 we can model the price of the asset as follows. The up and down movement of the price is described by a sequence of independent and identically distributed random variables $(Y_t)_{t=1,\dots,T}$ with distribution

$$\mathbb{P}(Y_1 = u) = p = 1 - \mathbb{P}(Y_1 = d)$$

for some probability $p \in (0, 1)$ and some $0 < d < 1 < u$, which implies no–arbitrage.

Consequently, the price process follows the recursion

$$S_0 \equiv 1, \quad S_{t+1} = S_t \cdot Y_{t+1}, \quad t = 0, 1, \dots, T-1.$$

The information is modelled by the natural filtration

$$\mathcal{F}_0 = \{\emptyset, \Omega\}, \quad \mathcal{F}_t := \sigma(Y_1, \dots, Y_t), \quad t = 1, \dots, T.$$

An investment strategy at time $t \in \{0, 1, \dots, T\}$ is a vector $\varphi_t := (b_t, a_t)$ modelling the amount of the current wealth that is invested in the bond and the asset respectively. This means that, for a given initial wealth $V_0 = b_0 + a_0$, the capital can be defined via

$$V_{t+1} := b_{t+1} + a_{t+1} = b_t + a_t Y_{t+1}, \quad t = 0, 1, \dots, T-1.$$

The last equation is the assumption that only self–financing strategies are admissible. We only consider these kind of strategies and furthermore assume that the strategy process $(\varphi_t)_{t=0,1,\dots,T}$ is $(\mathcal{F}_t)_{t=0,1,\dots,T}$–adapted. Proposition 5.7 in Föllmer and Schied (2004) yields that the self–financing property is equivalent to the wealth process fulfilling the recursion

$$V_{t+1}(a_t) := V_{t+1} = V_t + a_t(Y_{t+1} - 1), \quad t = 0, 1, \dots, T-1. \tag{2.1}$$

Hence, we can restrict ourselves to the decision about the amount a_t of the asset we want to hold. Our investment in the bond is then given by $b_t = V_t - a_t$. In general, a_t and b_t can be allowed to attain even negative values if we assume that shortselling is possible. We will discern these cases in the investigations below. Note that if shortselling is not allowed, we have to choose $a_t \in [0, V_t]$. We denote with F^T the set of all admissible policies as described above, see below for a more formal definition. If shortselling is not allowed, we write F^T_{ns} instead of F^T. Since the wealth process $(V_t)_{t=0,1,\dots,T}$ depends on the choice of a policy $\pi \in F^T$ it is sometimes denoted by $(V_t^\pi)_{t=0,1,\dots,T}$.

Let us now discuss the optimization problems described at the beginning. One objective is to maximize the expected utility of the final wealth under a risk constraint, i. e.

$$\begin{aligned} &\sup_{\pi \in F'} \mathbb{E}\left[u(V_T^\pi) \mid V_0 = v\right] \\ &\text{s.t. } \rho(V_T^\pi \mid V_0 = v) \le R \end{aligned} \tag{UM}$$

for some utility function $u : \mathbb{R} \to \mathbb{R}$, a risk measure ρ and a given maximal risk level $R \in \mathbb{R} \cup \{\infty\}$, where $R = \infty$ is interpreted as the unconstrained case. The set F' could mean either F^T or F^T_{ns}. A variant of this problem is to replace the risk constraint on the final value with intermediate constraints. This will be dealt with in Section 2.3. A second approach is to minimize the risk under the constraint that a certain minimal expected return can be achieved, i. e.

$$\begin{aligned}&\inf_{\pi \in F'} \rho(V_T^\pi \mid V_0 = v)\\&\text{s.t. } \mathbb{E}\left[V_T^\pi \mid V_0 = v\right] \geq \mu\end{aligned} \tag{RM}$$

for given $\mu \in \mathbb{R} \cup \{-\infty\}$. Hence, $\mu = -\infty$ represents the unconstrained case.

Here, we choose the identity function as a utility function, i. e. set $u := \mathrm{id}_{\mathbb{R}}$. This is a special case of a power utility function or HARA utility function as it is called in Föllmer and Schied (2004). As a risk measure, we incorporate Average Value at Risk, i. e. define $\rho := \mathrm{AV@R}_\gamma$ for given $\gamma \in (0,1)$. We have seen in the previous chapter that it is a reasonable choice in economically motivated models.

To solve these problems we have to introduce the theory of Markov decision processes. The following quantities are relevant for this approach.

- The *state space* is denoted by $S \subseteq \mathbb{R}$ and equipped with the σ–algebra $\mathcal{S} := \mathcal{B}_S$. Let $v \in S$ be an element of the state space, where v represents a realization of the wealth process at time $t \in \{0, 1, \dots, T\}$. If shortselling is not allowed we set $S := \mathbb{R}_+$, otherwise $S = \mathbb{R}$.
- The *action space* is $A = \mathbb{R}$ equipped with $\mathcal{A} := \mathcal{B}$. Then $a \in A$ denotes the invested amount in the asset.
- Sometimes there are restrictions on the actions, therefore for state $v \in S$ the space of *admissible policies* is $D(v) \subset A$ and the *restriction set* is then $D = \{(v,a) \in S \times A \mid a \in D(v)\}$. We assume that $D(v)$ is convex for $v \in S$. This is fulfilled in all of the treated models. For example, set $D(v) := [0, v]$ for $v \in S$ if $F' = F_{\mathrm{ns}}^T$ and if there are no intermediate constraints.
- The *disturbance* has values in $E = \{u, d\}$ equipped with $\mathcal{E} := \wp(E)$.
- The *transition function* $T_t : D \times E \to S$ at time $t = 1, \dots, T$ is given by

  $$T_t(v, a, y) := v + a \cdot (y - 1), \quad (v, a, y) \in D \times E.$$

- The *transition law* $Q_t : D \times \mathcal{S} \to [0, 1]$ at time $t = 1, \dots, T$ is the conditional distribution of V_t^π given $(V_{t-1}^\pi, a_{t-1}) = (v, a) \in D$, formally

  $$Q_t(v, a; B) := \mathbb{P}(T_t(v, a, Y_t) \in B), \quad B \in \mathcal{S}.$$

- The *one-step reward function* at time $t = 0, 1, \dots, T - 1$ is usually a measurable mapping $r_t : D \to \mathbb{R}$. In our models it vanishes, i. e. we can choose $r_t \equiv 0$, $t = 0, 1, \dots, T - 1$.
- The *terminal reward function* is a measurable mapping $J_T : S \to \mathbb{R}$. In the mean maximization problem without a constraint for example, it is chosen as $J_T(v) = v$, $v \in S$.

We are now ready to solve various optimization problems. The theoretical background is the following theorem for which we need another definition.

Definition 2.1. *For $t = 0, \dots, T-1$, the set of $(T-t)$–step* admissible Markov policies *is given by*

$$F^{T-t} := \{\pi = (f_t, \dots, f_{T-1}) \mid f_k : S \to A \ (\mathcal{S}, \mathcal{A})\text{–}meas., \ k = t, \dots, T-1\}.$$

By defining the classic value functions via

$$J_{t,\pi}(v) := \mathbb{E}\left[J_T(V_T^\pi) \mid V_t^\pi = v\right], \quad v \in S,$$

for every $\pi = (f_t, \dots, f_{T-1}) \in F^{T-t}$ and

$$J_t(v) := \inf_{\pi \in F^{T-t}} J_{t,\pi}(v), \quad v \in S,$$

Theorem 3.2.1 in Hernández-Lerma and Lasserre (1996) yields the following result.

Theorem 2.1. *Let $t \in \{0, 1, \dots, T-1\}$. If the functions $J_T'(v) := J_T(v)$ and*

$$J_t'(v) := \inf_{a \in D(v)} \mathbb{E}\left[J_{t+1}'(V_{t+1}) \,\middle|\, V_t = v, \, a_t = a\right], \quad v \in S, \tag{2.2}$$

defined for $t = 0, \dots, T-1$ are measurable and if the infimum is attained in $a^* = f_t^*(v) \in D(v)$, such that $f_t^* : S \to A$ is a $(\mathcal{S}, \mathcal{A})$–measurable function, then $\pi^* := (f_0^*, \dots, f_{T-1}^*) \in F^T$ is an optimal policy in the sense that

$$J_0(v) = J_{0,\pi^*}(v) = J_0'(v), \quad v \in S.$$

2.2 Risk minimization

In this section we want to solve the risk minimization problem (RM) introduced above with $\rho := \text{AV@R}_\gamma$ for given $\gamma \in (0,1)$. We assume throughout this section that shortselling is possible, i. e. choose $F' = F^T$. Furthermore, define the probability

$$p^* := \frac{1-d}{u-d} \in (0,1).$$

Note that we have

$$p > p^* \iff \mathbb{E}[Y_1 - 1] > 0, \tag{2.3}$$

i. e. there is a tendency for an increase in (2.1) if $p > p^*$ and a decrease if $p < p^*$ (if we invest a positive amount in the asset). Hence, we can interpret our problem as a favorable game in the former case and as an unfavorable game in the latter case. Note that p^* is the risk–neutral probability in the CRRM, i. e. it defines the unique martingale measure $\mathbb{P}^*$ in this model via $\mathbb{P}^*(Y_t = u) = p^*$, $t = 1, \dots, T$, see e. g. Theorem 5.40 in Föllmer and Schied (2004).

2.2.1 The unconstrained case

As an introductory step, let us first solve (RM) without a constraint. Formally, we set $\mu = -\infty$. Below, we discern the cases $p > p^*$ and $p < p^*$. For the sake of completeness, let us briefly make some statements if $p = p^*$.

Lemma 2.1. *Assume $p = p^*$ and let $v \in S$, $\gamma \in (0,1)$. Then the following holds for arbitrary $\pi \in F^T$:*

$$\begin{aligned} &(i) \quad \mathbb{E}\left[V_T^\pi \mid V_0 = v\right] = v, \\ &(ii) \quad \mathrm{AV@R}_\gamma(V_T^\pi \mid V_0 = v) \geq -v. \end{aligned}$$

Proof. In the case $p = p^*$, $(V_t^\pi)_{t=0,1,\dots,T}$ is a $\mathbb{P}$-martingale for every $\pi \in F^T$. Hence, $\mathbb{E}\left[V_T^\pi \mid V_0 = v\right] = v$. Furthermore, it holds for any random variable X that $\mathrm{AV@R}_\gamma(X) \geq \mathbb{E}\left[-X\right]$, therefore yielding also the second part. □

Now, the following result holds.

Proposition 2.1. *Let $\gamma \in (0,1)$ and $v \in S$. Then it holds*

$$\inf_{\pi \in F^T} \mathrm{AV@R}_\gamma(V_T^\pi \mid V_0 = v) = \begin{cases} -v & , \quad \gamma \geq 1 - \left(\min\left\{\frac{1-p}{1-p^*}, \frac{p}{p^*}\right\}\right)^T, \\ -\infty & , \quad \textit{otherwise.} \end{cases}$$

In the finite case, an optimal policy is given by $\pi^* = (0, \dots, 0)$.

Proof. If $p = p^*$ the assertion is obvious using Lemma 2.1 and noting that the lower bound $-v$ given in part (ii) is attained for $\pi = (0, \dots, 0)$. Now, assume $p \neq p^*$. The representation (1.5) yields

$$\begin{aligned} &\inf_{\pi \in F^T} \mathrm{AV@R}_\gamma(V_T^\pi \mid V_0 = v) \\ &= \inf_{\pi \in F^T} \inf_{b \in \mathbb{R}} \left\{ b + \frac{1}{1-\gamma} \mathbb{E}\left[(-V_T^\pi - b)^+ \mid V_0 = v\right] \right\} \\ &= \inf_{b \in \mathbb{R}} \left\{ b + \frac{1}{1-\gamma} \inf_{\pi \in F^T} \mathbb{E}\left[(-V_T^\pi - b)^+ \mid V_0 = v\right] \right\} \\ &= \begin{cases} \inf_{b \in \mathbb{R}} \left\{ b + \frac{1}{1-\gamma} \left(\frac{1-p}{1-p^*}\right)^T (-v-b)^+ \right\} & , \quad p > p^*, \\ \inf_{b \in \mathbb{R}} \left\{ b + \frac{1}{1-\gamma} \left(\frac{p}{p^*}\right)^T (-v-b)^+ \right\} & , \quad p < p^*, \end{cases} \end{aligned}$$

where the last step follows from Theorem 4.1 in Runggaldier et al. (2002). If we set

$$c := \begin{cases} \frac{1}{1-\gamma} \left(\frac{1-p}{1-p^*}\right)^T & , \quad p > p^*, \\ \frac{1}{1-\gamma} \left(\frac{p}{p^*}\right)^T & , \quad p < p^*, \end{cases}$$

then the function

$$b \mapsto b + c \cdot (-v - b)^+, \quad b \in \mathbb{R},$$

is linearly increasing for $b > -v$. Furthermore, it is non-increasing for $b < -v$ if and only if $c \geq 1$. In this case, the minimal value is attained in $b^* = -v$. Otherwise, i. e. if $c < 1$, the function is strictly increasing and piecewise linear on the real line, hence tending to $-\infty$ for $b \to -\infty$.

Since

$$c \geq 1 \Leftrightarrow \begin{cases} \gamma \geq 1 - \left(\frac{1-p}{1-p^*}\right)^T & , \quad p > p^* \\ \gamma \geq 1 - \left(\frac{p}{p^*}\right)^T & , \quad p < p^*, \end{cases}$$

the assertion follows. Note that optimality of the policy π^* in the finite case follows because $V_T^{\pi^*} = V_0 \in \mathbb{R}$ and $\mathrm{AV@R}_\gamma(V_0) = -V_0$. □

Remark. 1. The optimal policy is not unique. We can observe from the proof and Theorem 4.1 in Runggaldier et al. (2002) that for $p > p^*$, for example, also $\pi = (f_0, \dots, f_{T-1})$ with

$$f_t(v) = -\frac{v - V_0}{u - 1}, \quad v \in S, t = 0, 1, \dots, T - 1,$$

is optimal. To apply the mentioned Theorem, we incorporated a constant claim $H(S_T) := -b$.

2. It is important to observe that exactly one of the two values $\frac{1-p}{1-p^*}$ and $\frac{p}{p^*}$ is less than 1 while the other one is greater than 1, depending on whether $p > p^*$ or $p < p^*$. Hence, the inequality for γ in the proposition is always non-trivial.

3. We see by the result that if the safety level γ is chosen very small, i. e. we are a risk-lover, it is possible to choose a policy π which yields an arbitrarily small risk. However, it is more reasonable to choose γ large, hence our minimal risk will usually be larger than $-\infty$. This interpretation also applies to the results given below.

Interpretation of the optimal policy The proposition shows that if we choose a sufficiently large safety level γ we minimize our risk by investing nothing in the risky asset and obtain a minimal risk which is equal to $\mathrm{AV@R}_\gamma(V_0 \mid V_0 = v)$. Hence, it is not possible to diminish this initial risk further by investing a positive proportion of the wealth into the asset over time.

2.2.2 A constraint on the final value

Now, we want to solve (RM) with general $\mu \in \mathbb{R}$. As a preparation, we need a simple lemma. Set $Y := Y_1$. Then we have by a direct calculation for $\gamma \in (0,1)$

$$\begin{aligned} \mathrm{AV@R}_\gamma(Y) &= -d - \mathbb{1}_{[0,p]}(\gamma)\frac{p-\gamma}{1-\gamma}(u-d) \leq -d, \\ \mathrm{AV@R}_\gamma(-Y) &= u + \mathbb{1}_{[0,1-p]}(\gamma)\frac{1-p-\gamma}{1-\gamma}(d-u) \leq u. \end{aligned} \tag{2.4}$$

The relationship of p and p^* now provides some properties of these two values.

Lemma 2.2. *Let* $\gamma \in (0,1)$.

(i) If $p > p^*$, *then* $\mathrm{AV@R}_\gamma(-Y) > 1$. *Moreover,* $p > \frac{p-p^*}{1-p^*}$ *and*

$$\mathrm{AV@R}_\gamma(Y) \begin{cases} > -1 & , \quad \gamma > \dfrac{p-p^*}{1-p^*}, \\ = -1 & , \quad \gamma = \dfrac{p-p^*}{1-p^*}, \\ < -1 & , \quad \gamma < \dfrac{p-p^*}{1-p^*}. \end{cases}$$

(ii) If $p < p^*$, *then* $\mathrm{AV@R}_\gamma(Y) > -1$. *Moreover,* $1-p > \frac{p^*-p}{p^*}$ *and*

$$\mathrm{AV@R}_\gamma(-Y) \begin{cases} > 1 & , \quad \gamma > \dfrac{p^*-p}{p^*}, \\ = 1 & , \quad \gamma = \dfrac{p^*-p}{p^*}, \\ < 1 & , \quad \gamma < \dfrac{p^*-p}{p^*}. \end{cases}$$

Proof. The proof is given for part (i). Part (ii) can be done in a similar manner. Note that the inequalities $p > \frac{p-p^*}{1-p^*}$ and $1-p > \frac{p^*-p}{p^*}$ are obvious.

Let $p > p^*$. For every random variable X it holds $\mathrm{AV@R}_\gamma(X) \geq \mathbb{E}[-X]$. Hence,

$$\mathrm{AV@R}_\gamma(-Y) \geq \mathbb{E}[Y] \overset{(2.3)}{>} 1.$$

Therefore the first part of (i) follows.

Let $\gamma \geq p > \frac{p-p^*}{1-p^*}$. We obtain $\mathrm{AV@R}_\gamma(Y) = -d > -1$. Consider now the converse $\gamma < p$. Then the following holds:

$$\begin{aligned} \mathrm{AV@R}_\gamma(Y) > -1 \;&\Leftrightarrow\; -d - \frac{p-\gamma}{1-\gamma}(u-d) > -1 \\ &\Leftrightarrow\; 1-d > \frac{p-\gamma}{1-\gamma}(u-d) \\ &\Leftrightarrow\; p^* > \frac{p-\gamma}{1-\gamma} \\ &\Leftrightarrow\; \gamma > \frac{p-p^*}{1-p^*}. \end{aligned}$$

Doing the same calculation replacing ">" with "=" and "<" finally yields the assertion. □

As an illustration for the solution of problem (RM) let us first take a look at the one–period case. This can be treated in an elementary way.

The one–period case

Let $T = 1$ and $v \in S$ be given and assume $p > p^*$. We have to solve

$$\begin{aligned}&\inf_{a\in\mathbb{R}} \mathrm{AV@R}_\gamma(v + a(Y_1 - 1))\\ &\text{s.t. } \mathbb{E}\left[v + a(Y_1 - 1)\right] \geq \mu.\end{aligned}$$

Denote the optimal target value by $J_0(v)$. The constraint can be reformulated with $a \in \mathbb{R}$ as

$$\mathbb{E}\left[v + a(Y_1 - 1)\right] \geq \mu \;\Leftrightarrow\; a \geq \frac{\mu - v}{\mathbb{E}\left[Y_1 - 1\right]}.$$

If $v \leq \mu$, only non–negative values for a are admissible. Hence, the target function becomes by coherence of Average Value–at–Risk

$$h(a) := \mathrm{AV@R}_\gamma(v + a(Y_1 - 1)) = a(\mathrm{AV@R}_\gamma(Y_1) + 1) - v, \quad a \in \mathbb{R}_+.$$

By Lemma 2.2 (i), the function h is non–decreasing on $\mathbb{R}_+$ if and only if $\gamma \geq \frac{p-p^*}{1-p^*}$. In this case, the optimal value for a is the smallest admissible one, i. e. we have $a^* = \frac{\mu-v}{\mathbb{E}[Y_1-1]}$. It is unique if and only if $\gamma > \frac{p-p^*}{1-p^*}$. If $\gamma < \frac{p-p^*}{1-p^*}$, h is linear and strictly decreasing, hence tending to $-\infty$ for $a \to \infty$.

If $v > \mu$, also negative values for a are admissible. We obtain

$$h(a) = -a(\mathrm{AV@R}_\gamma(-Y_1) - 1) - v, \quad a \in \mathbb{R}_-,$$

and the function h is decreasing on $\mathbb{R}_-$ again by Lemma 2.2 (i). With the same argument as above the optimal value for a is $a^* = 0$ if $\gamma \geq \frac{p-p^*}{1-p^*}$ and h tends to $-\infty$ otherwise. Summing things up, we have

$$J_0(v) = \begin{cases} (\mu - v)^+ \cdot \dfrac{\mathrm{AV@R}_\gamma(Y_1) + 1}{\mathbb{E}\left[Y_1 - 1\right]} - v &, \quad \gamma \geq \dfrac{p - p^*}{1 - p^*}, \\ -\infty &, \quad \gamma < \dfrac{p - p^*}{1 - p^*}. \end{cases}$$

In the same manner we obtain for $p < p^*$

$$J_0(v) = \begin{cases} (\mu - v)^+ \cdot \dfrac{\mathrm{AV@R}_\gamma(-Y_1) - 1}{\mathbb{E}\left[Y_1 - 1\right]} - v &, \quad \gamma \geq \dfrac{p^* - p}{p^*}, \\ -\infty &, \quad \gamma < \dfrac{p^* - p}{p^*}. \end{cases}$$

If $p = p^*$, no policy is admissible for $v < \mu$ and all policies are admissible otherwise. In the latter case, $a = 0$ is optimal by the lower bound given in Lemma 2.1 (ii). Hence the representations for $p > p^*$ are also valid for $p = p^*$ if we set $0 \cdot \infty = 0$.

The multi–period case

In order to treat this problem, we take a Lagrangian approach. Define the Lagrange function $L : F^T \times \mathbb{R}_+ \to \mathbb{R}$ via

$$L(\pi, \lambda) := \mathrm{AV@R}_\gamma(V_T^\pi \mid V_0 = v) + \lambda\big(\mu - \mathbb{E}\,[V_T^\pi \mid V_0 = v]\big), \quad (\pi, \lambda) \in F^T \times \mathbb{R}_+.$$

The optimal value of (RM) is then given by

$$h_1(v) := \inf_{\pi \in F^T} \sup_{\lambda \geq 0} L(\pi, \lambda), \quad v \in S. \tag{2.5}$$

To find an optimal pair (π^*, λ^*) for (RM) we first solve the *dual problem*

$$h_2(v) := \sup_{\lambda \geq 0} \inf_{\pi \in F^T} L(\pi, \lambda), \quad v \in S, \tag{2.6}$$

which in general gives only a lower bound for the *primal problem* (RM), i. e. it holds $h_1 \geq h_2$. However, we will see below that the optimal target values are equal and how we obtain an optimal policy for (RM) from an optimal pair (π^*, λ^*) of the dual problem (2.6).

The latter can be rewritten as follows, again applying (1.5):

$$\begin{aligned}
&h_2(v) \\
&= \sup_{\lambda \geq 0} \inf_{\pi \in F^T} \Big\{ \mathrm{AV@R}_\gamma(V_T^\pi \mid V_0 = v) + \lambda\big(\mu - \mathbb{E}\,[V_T^\pi \mid V_0 = v]\big) \Big\} \\
&= \sup_{\lambda \geq 0} \inf_{\pi \in F^T} \Big\{ \inf_{b \in \mathbb{R}} \Big\{ b + \frac{1}{1-\gamma} \mathbb{E}\,\big[(-V_T^\pi - b)^+ \mid V_0 = v\big] \Big\} + \lambda\big(\mu - \mathbb{E}\,[V_T^\pi \mid V_0 = v]\big) \Big\} \\
&= \sup_{\lambda \geq 0} \inf_{b \in \mathbb{R}} \Big\{ b(1 + \lambda) + \lambda\mu + \inf_{\pi \in F^T} \Big\{ \Big(\lambda + \frac{1}{1-\gamma}\Big) \mathbb{E}\,\big[(-V_T^\pi - b)^+ \mid V_0 = v\big] \\
&\qquad\qquad - \lambda \mathbb{E}\,\big[(-V_T^\pi - b)^- \mid V_0 = v\big] \Big\}\Big\} \\
&= \sup_{\lambda \geq 0} \inf_{b \in \mathbb{R}} \Big\{ b(1 + \lambda) + \lambda\mu + \inf_{\pi \in F^T} \mathbb{E}\,\big[J_T(V_T^\pi) \mid V_0 = v\big] \Big\},
\end{aligned}$$

where we used $V_T^\pi + b = (V_T^\pi + b)^+ - (V_T^\pi + b)^-$, $b \in \mathbb{R}$, for $\pi \in F^T$ and set

$$J_T(v') = \Big(\lambda + \frac{1}{1-\gamma}\Big)(v' + b)^- - \lambda(v' + b)^+, \quad v' \in S.$$

Consequently, as a first step we have to solve the Markov decision model introduced above with J_T as defined to obtain the inner infimum. Again, we give a detailed proof in the case $p > p^*$ and only state the similar result if $p < p^*$. Define one step policies for $b \in \mathbb{R}$ via

$$f_b^{(1)}(v') := -\frac{v' + b}{u - 1}, \quad f_b^{(2)}(v') := \frac{v' + b}{1 - d}, \quad v' \in S.$$

The theorem is formulated for a slightly more general form of J_T.

Theorem 2.2. *Let $\alpha_1 > \alpha_2 \geq 0$, $b \in \mathbb{R}$, and assume $p \neq p^*$. Furthermore, define $J_T(v) := \alpha_1(v+b)^- - \alpha_2(v+b)^+$ and $D(v) = A$, $v \in S$. Then it holds:*

(i) If $p > p^*$ we have for $v \in S$

$$J_0(v) = \begin{cases} \left(\frac{1-p}{1-p^*}\right)^T \alpha_1(v+b)^- - \left(\frac{p}{p^*}\right)^T \alpha_2(v+b)^+ , & \left(\frac{1-p}{1-p^*}\right)^T \alpha_1 \geq \left(\frac{p}{p^*}\right)^T \alpha_2, \\ -\infty , & \textit{otherwise.} \end{cases}$$

In the finite case, an optimal policy $\pi_b = (f_0, \dots, f_{T-1}) \in F^T$ is given by

$$f_t = \max\left\{f_b^{(1)}, f_b^{(2)}\right\}, \quad t = 0, 1, \dots, T-1.$$

(ii) If $p < p^*$ we have for $v \in S$

$$J_0(v) = \begin{cases} \left(\frac{p}{p^*}\right)^T \alpha_1(v+b)^- - \left(\frac{1-p}{1-p^*}\right)^T \alpha_2(v+b)^+ , & \left(\frac{p}{p^*}\right)^T \alpha_1 \geq \left(\frac{1-p}{1-p^*}\right)^T \alpha_2, \\ -\infty , & \textit{otherwise.} \end{cases}$$

In the finite case, an optimal policy $\pi_b = (f_0, \dots, f_{T-1}) \in F^T$ is given by

$$f_t = \min\left\{f_b^{(1)}, f_b^{(2)}\right\}, \quad t = 0, 1, \dots, T-1.$$

Proof. We only show part (i), so assume $p > p^*$. If we can prove that for all $t \in \{0, 1, \dots, T-1\}$ and $v \in S$

$$J_t'(v) = \begin{cases} \left(\frac{1-p}{1-p^*}\right)^{T-t} \alpha_1(v+b)^- - \left(\frac{p}{p^*}\right)^{T-t} \alpha_2(v+b)^+ , & \left(\frac{1-p}{1-p^*}\right)^{T-t} \alpha_1 \geq \left(\frac{p}{p^*}\right)^{T-t} \alpha_2, \\ -\infty , & \text{otherwise,} \end{cases}$$

the assertion will follow from Theorem 2.1.

The proof is by backward induction on t and the main work has to be done for $t = T-1$. By definition of J_{T-1}', we have to solve for $v \in S$

$$\begin{aligned} J_{T-1}'(v) &= \inf_{a \in A} \mathbb{E}\left[\alpha_1(v+b+a(Y_T-1))^- - \alpha_2(v+b+a(Y_T-1))^+\right] \\ &= \inf_{a \in A} \Big\{ p\big(\alpha_1(v+b+a(u-1))^- - \alpha_2(v+b+a(u-1))^+\big) \\ &\qquad + (1-p)\big(\alpha_1(v+b+a(d-1))^- - \alpha_2(v+b+a(d-1))^+\big)\Big\}. \end{aligned}$$

Assume first that $v \geq -b$. In this case

$$\psi^{(1)} := f_b^{(1)}(v) = -\frac{v+b}{u-1} \leq 0 \leq -\frac{v+b}{d-1} = f_b^{(2)}(v) =: \psi^{(2)}.$$

We have to analyze the slope of the piecewise linear function from above – which has to be minimized – on the three intervals $(-\infty, \psi^{(1)})$, $[\psi^{(1)}, \psi^{(2)}]$ and $(\psi^{(2)}, \infty)$. For $a \in (\psi^{(2)}, \infty)$, we have $v+b+a(u-1) > 0$, $v+b+a(d-1) < 0$ and the slope becomes negative on this interval if and only if

$$\begin{aligned} & -p\alpha_2(u-1) - (1-p)\alpha_1(d-1) < 0 \\ \Leftrightarrow\quad & -p\alpha_2\frac{u-1}{u-d} - (1-p)\alpha_1\frac{d-1}{u-d} < 0 \\ \Leftrightarrow\quad & \frac{1-p}{1-p^*}\alpha_1 < \frac{p}{p^*}\alpha_2. \end{aligned} \tag{2.7}$$

Hence, by choosing a arbitrarily large, we obtain $J'_{T-1}(v) = -\infty$ if (2.7) holds.

Assume now that $\frac{1-p}{1-p^*}\alpha_1 \geq \frac{p}{p^*}\alpha_2$ so that there is an optimal value for a with $a \leq \psi^{(2)}$ (note that if equality holds in (2.7), also every $a > \psi^{(2)}$ is optimal). If even $a < \psi^{(1)}$ then $v+b+a(u-1) < 0$, $v+b+a(d-1) > 0$ and the slope on $(-\infty, \psi^{(1)})$ is always negative because

$$\begin{aligned} & -p\alpha_1(u-1) - (1-p)\alpha_2(d-1) < 0 \\ \Leftrightarrow\quad & -p\alpha_1\frac{u-1}{u-d} - (1-p)\alpha_2\frac{d-1}{u-d} < 0 \\ \Leftrightarrow\quad & \frac{p^*}{p}\alpha_2 < \frac{1-p^*}{1-p}\alpha_1, \end{aligned}$$

which is always fulfilled because $\alpha_1 > \alpha_2$ and $p > p^*$. Consequently, an optimal value a^* for a fulfills $a^* \geq \psi^{(1)}$, hence $a^* \in [\psi^{(1)}, \psi^{(2)}]$. The last step is to analyze the slope on this interval. We obtain $v+b+a(u-1) \geq 0$, $v+b+a(d-1) \geq 0$ for $a \in [\psi^{(1)}, \psi^{(2)}]$ and therefore the function is non-increasing on this region:

$$\begin{aligned} & -p\alpha_2(u-1) - (1-p)\alpha_2(d-1) \leq 0 \\ \Leftrightarrow\quad & -\alpha_2\Big(p\frac{u-1}{u-d} + (1-p)\frac{d-1}{u-d}\Big) \leq 0 \\ \Leftrightarrow\quad & \alpha_2\underbrace{(p(1-p^*) - (1-p)p^*)}_{>0 \text{ for } p>p^*} \geq 0. \end{aligned}$$

This always holds since by assumption $\alpha_2 \geq 0$. We conclude that a (not necessarily unique) optimal value for a is given by $a^* = \psi^{(2)}$ and that

$$\begin{aligned} J'_{T-1}(v) &= -p\alpha_2(v+b+\psi^{(2)}(u-1)) = -p\alpha_2\Big(v+b-(v+b)\frac{u-1}{d-1}\Big) \\ &= -\frac{p}{p^*}\alpha_2(v+b), \quad v \geq -b. \end{aligned} \tag{2.8}$$

Now, let $v < -b$. We then have $\psi^{(2)} < \psi^{(1)}$ and we can show in a similar way as above that the function is strictly decreasing on $(\psi^{(1)}, \infty)$ if and only if (2.7) holds, hence again $J'_{T-1}(v) = -\infty$. Assuming the converse $\frac{1-p}{1-p^*}\alpha_1 \geq \frac{p}{p^*}\alpha_2$ shows that the infimum is attained in $a^* = \psi^{(1)}$ which yields

$$J'_{T-1}(v) = -(1-p)\alpha_1(v + b + \psi^{(1)}(d-1)) = -\frac{1-p}{1-p^*}\alpha_2(v+b), \quad v < -b. \tag{2.9}$$

Combining (2.8) and (2.9) gives the desired result for $t = T-1$. The optimal policy f_{T-1} is given by

$$f_{T-1}(v) = \max\{\psi^{(1)}, \psi^{(2)}\} = \max\left\{f_b^{(1)}(v), f_b^{(2)}(v)\right\},, \quad v \in S,$$

which is indeed an $(\mathcal{S}, \mathcal{A})$-measurable function.

Finally, assume that the assertion holds for fixed $t \in \{1, \ldots, T-1\}$. If

$$\left(\frac{1-p}{1-p^*}\right)^{T-t} \alpha_1 > \left(\frac{p}{p^*}\right)^{T-t} \alpha_2$$

the assertion follows from the induction hypothesis exactly in the same way as for $t = T-1$ by replacing α_1 with $\left(\frac{1-p}{1-p^*}\right)^{T-t}\alpha_1$ and α_2 with $\left(\frac{p}{p^*}\right)^{T-t}\alpha_2$. Now, the converse

$$C := \left(\frac{1-p}{1-p^*}\right)^{T-t} \alpha_1 \leq \left(\frac{p}{p^*}\right)^{T-t} \alpha_2, \tag{2.10}$$

implies

$$\left(\frac{1-p}{1-p^*}\right)^{T-t+1} \alpha_1 < \left(\frac{p}{p^*}\right)^{T-t+1} \alpha_2.$$

Hence, we have to show $J'_{t-1}(v) = -\infty$ for all $v \in S$. If the inequality in (2.10) is strict, we obtain $J'_t(v) = -\infty$ for all $v \in S$, hence also $J'_{t-1}(v) = -\infty$ for all $v \in S$. In the case of equality in (2.10) it holds for the constant $C > 0$ and $v \in S$

$$\begin{aligned} J'_{t-1}(v) &= \inf_{a \in A} \mathbb{E}\left[J'_t(V_t) \mid V_{t-1} = v, a_{t-1} = a\right] \\ &= C \cdot \inf_{a \in \mathbb{R}} \mathbb{E}\left[(v + b + a(Y_t - 1))^- - (v + b + a(Y_t - 1))^+\right] \\ &= -C \cdot \sup_{a \in \mathbb{R}} \left\{v + b + a(\mathbb{E}\left[Y_t - 1\right])\right\} = -\infty. \end{aligned}$$

The last equality follows because $p \neq p^*$. This concludes the proof. □

Remark. 1. One can observe from the proof that π_b is not always the unique optimal policy. First, if $\alpha_2 = 0$ additionally every f_t with

$$f_t(v) \in \left[\min\left\{f_b^{(1)}(v), f_b^{(2)}(v)\right\}, \max\left\{f_b^{(1)}(v), f_b^{(2)}(v)\right\}\right]$$

is optimal for all $t \in \{0, 1, \ldots, T-1\}$. If on the other hand $\alpha_2 > 0$ and if

$$\left(\frac{1-p}{1-p^*}\right)^T \alpha_1 = \left(\frac{p}{p^*}\right)^T \alpha_2 \text{ respectively } \left(\frac{p}{p^*}\right)^T \alpha_1 = \left(\frac{1-p}{1-p^*}\right)^T \alpha_2$$

then every f_0 with $f_0(v) \geq \max\left\{f_b^{(1)}(v), f_b^{(2)}(v)\right\}$ (for $p > p^*$) or with $f_0(v) \leq \min\left\{f_b^{(1)}(v), f_b^{(2)}(v)\right\}$ (for $p < p^*$ respectively) is also optimal. The components $f_1, \dots, f_{T-1}$ of π_b are unique though.

2. Theorem 4.1 in Runggaldier et al. (2002) is a special case of this result with $\alpha_1 = 1$ and $\alpha_2 = 0$. In order to obtain the same optimal policies we refer to the first part of this remark. Furthermore, our proof still works with an analogous result if one replaces b with a claim to be hedged as they do and by extending the state space.

3. It is not possible to use Theorem 3.2 from Favero and Vargiolu (2006) with $l : \mathbb{R} \to \mathbb{R}$, $l(x) := \alpha_1 x^- - \alpha_2 x^+$, and a claim $H(S_T) \equiv b$ for example since an important assumption would be that $V_0 < b$. But in our model, it is not reasonable to exclude an initial wealth smaller than b.

To obtain optimal policies for (RM) we need the following lemma, which provides the expectation of the final wealth under the policies given in Theorem 2.2.

Lemma 2.3. *Let $b \in \mathbb{R}$ and consider the policy π_b defined in Theorem 2.2. Then it holds for $v \in S$:*

(i) If $p > p^*$ and $b \geq -v$ or if $p < p^*$ and $b \leq -v$ it holds

$$\mathbb{E}\left[V_T^{\pi_b} \mid V_0 = v\right] = v + (v+b) \cdot \frac{p^T - (p^*)^T}{(p^*)^T} \geq v.$$

(ii) If $p > p^*$ and $b \leq -v$ or if $p < p^*$ and $b \geq -v$ it holds

$$\mathbb{E}\left[V_T^{\pi_b} \mid V_0 = v\right] = v - (v+b) \cdot \frac{(1-p^*)^T - (1-p)^T}{(1-p^*)^T} \geq v.$$

Proof. We only prove the case $p > p^*$ and $b \geq -v$. Let $\pi_b = (f_0, \dots, f_{T-1})$. First, a simple inductive argument yields $V_t^{\pi_b} + b \geq 0$ for every $t \in \{0, 1, \dots, T-1\}$ if $V_0 = v$. Hence, $f_t(V_t^{\pi_b}) = \frac{V_t^{\pi_b} + b}{1-d}$. We furthermore obtain

$$V_t^{\pi_b} = v + (v+b) \cdot Z_t, \quad t = 0, 1, \dots, T, \tag{2.11}$$

where $Z = (Z_t)_{t=0,1,\dots,T}$ is a non-negative Markov chain defined recursively through

$$Z_0 = 0, \; Z_t = Z_{t-1} + \frac{1 + Z_{t-1}}{1-d}(Y_t - 1), \quad t = 1, \dots, T.$$

This can be seen inductively. The case $t = 0$ is obvious. Now assume that the assertion holds for $V_{t-1}^{\pi_b}$. Then

$$\begin{aligned} V_t^{\pi_b} &= V_{t-1}^{\pi_b} + f_{t-1}(V_{t-1}^{\pi_b})(Y_t - 1) = V_{t-1}^{\pi_b} + \frac{V_{t-1}^{\pi_b} + b}{1-d}(Y_t - 1) \\ &= v + (v+b) \cdot Z_{t-1} + \frac{v + (v+b) \cdot Z_{t-1} + b}{1-d}(Y_t - 1) = v + (v+b) \cdot Z_t, \end{aligned}$$

hence (2.11) is shown. By construction, the expectations of the components of Z follow the recursion

$$\mathbb{E}\, Z_0 = 0,\ \mathbb{E}\, Z_t = \mathbb{E}\, Z_{t-1} + \frac{1+\mathbb{E}\, Z_{t-1}}{1-d}\mathbb{E}\,[Y_t - 1] = \frac{p-p^*}{p^*} + \mathbb{E}\, Z_{t-1}\cdot\frac{p}{p^*},\ t = 1,\dots,T,$$

hence

$$\mathbb{E}\, Z_t = \frac{p^t - (p^*)^t}{(p^*)^t}, \quad t = 0, 1, \dots, T. \tag{2.12}$$

We conclude that for every π_b with $b \geq -v$

$$\mathbb{E}\,[V_T^{\pi_b} \mid V_0 = v] \overset{(2.11)}{=} v + (v+b)\cdot \mathbb{E}\, Z_T \overset{(2.12)}{=} v + (v+b)\cdot \frac{p^T - (p^*)^T}{(p^*)^T}.$$

□

Now consider Theorem 2.2 with $\alpha_1 := \lambda + (1-\gamma)^{-1}$ and $\alpha_2 = \lambda$ for fixed $\lambda \geq 0$. To stress the dependence of the value function from λ and b set $J_0^{\lambda,b} := J_0$. Define

$$\gamma_1(p,p^*) := \begin{cases} \dfrac{(1-p^*)^T - (1-p)^T}{(1-p^*)^T} & ,\ p > p^*, \\ 0 & ,\ p = p^*, \\ \dfrac{(p^*)^T - p^T}{(p^*)^T} & ,\ p < p^*, \end{cases}$$

$$\gamma_2(p,p^*) := \begin{cases} p^T \dfrac{(1-p^*)^T - (1-p)^T}{\big(p(1-p^*)\big)^T - \big((1-p)p^*\big)^T} & ,\ p > p^*, \\ 0 & ,\ p = p^*, \\ (1-p)^T \dfrac{(p^*)^T - p^T}{\big((1-p)p^*\big)^T - \big(p(1-p^*)\big)^T} & ,\ p < p^*, \end{cases}$$

and

$$C_\gamma(p,p^*) := \begin{cases} \min\left\{ \dfrac{(p^*)^T}{p^T - (p^*)^T}, \dfrac{(1-p)^T(1-\gamma)^{-1} - (1-p^*)^T}{(1-p^*)^T - (1-p)^T} \right\} & ,\ p > p^*, \\ \infty & ,\ p = p^*, \\ \min\left\{ \dfrac{(1-p^*)^T}{(1-p)^T - (1-p^*)^T}, \dfrac{p^T(1-\gamma)^{-1} - (p^*)^T}{(p^*)^T - p^T} \right\} & ,\ p < p^*. \end{cases}$$

Note that always $\gamma_2(p,p^*) \geq \gamma_1(p,p^*)$ and that under the condition $\gamma \geq \gamma_1(p,p^*)$, we have $C_\gamma(p,p^*) \geq 0$. The following lemma provides the solution of the inner infimum in (2.6).

Lemma 2.4. *Let $\lambda \geq 0$ and assume $p \neq p^*$. Then we have with $V_0 = v \in S$*

$$\inf_{\pi \in F^T} L(\pi,\lambda) = \begin{cases} (\mu - v)\lambda - v & ,\ \lambda \in [0, C_\gamma(p,p^*)],\ \gamma \geq \gamma_1(p,p^*), \\ -\infty & ,\ \textit{otherwise}. \end{cases}$$

In the finite case, the unique optimal policy for $0 < \lambda < C_\gamma(p,p^*)$ is given by $\pi^* = \pi_{-v}$, where π_b is defined as in Theorem 2.2 for $b \in \mathbb{R}$. For $\lambda = 0$ the same policy is optimal but not unique, see the remark after Theorem 2.2 for $\alpha_2 = 0$.

If $\lambda = C_\gamma(p,p^*) > 0$ the set of optimal policies is given as follows. In the case $\gamma \geq \gamma_2(p,p^*)$ every π_b with $b \geq -v$ is optimal. For $\gamma_1(p,p^*) \leq \gamma \leq \gamma_2(p,p^*)$ one has to choose $b \leq -v$.

Proof. We have seen above that

$$\begin{aligned}
&\inf_{\pi\in F^T} L(\pi,\lambda)\\
&= \inf_{\pi\in F^T}\inf_{b\in\mathbb{R}}\left\{b(1+\lambda)+\lambda\mu+\mathbb{E}\left[\left(\lambda+\frac{1}{1-\gamma}\right)(V_T^\pi+b)^- - \lambda(V_T^\pi+b)^+ \,|\, V_0=v\right]\right\}\\
&= \inf_{b\in\mathbb{R}}\left\{b(1+\lambda)+\lambda\mu+\inf_{\pi\in F^T}\mathbb{E}\left[\left(\lambda+\frac{1}{1-\gamma}\right)(V_T^\pi+b)^- - \lambda(V_T^\pi+b)^+ \,|\, V_0=v\right]\right\}\\
&= \inf_{b\in\mathbb{R}}\{\underbrace{b(1+\lambda)+\lambda\mu+J_0^{\lambda,b}(v)}_{=:\,g(\lambda,b)}\},
\end{aligned}$$

where in the last step we used Theorem 2.2 with $\alpha_1 = \lambda + (1-\gamma)^{-1}$ and $\alpha_2 = \lambda$ for fixed $\lambda \geq 0$. The optimal policy for the inner infimum is given by π_b defined in the theorem for fixed $b \in \mathbb{R}$. Let $p > p^*$. Again by Theorem 2.2 we have for $b \in \mathbb{R}$ and $\lambda \geq 0$

$$\begin{aligned}
J_0^{\lambda,b}(v) = -\infty \;&\Leftrightarrow\; \left(\frac{1-p}{1-p^*}\right)^T(\lambda+(1-\gamma)^{-1}) < \left(\frac{p}{p^*}\right)^T\lambda\\
&\Leftrightarrow\; \lambda > \frac{\left((1-p)p^*\right)^T}{\left(p(1-p^*)\right)^T-\left((1-p)p^*\right)^T}(1-\gamma)^{-1}.
\end{aligned}$$

Consequently, $\inf_{\pi\in F^T} L(\pi,\lambda) = -\infty$ for all $\lambda \in \mathbb{R}_+ \setminus I_1$, where

$$I_1 := \left[0, \frac{\left((1-p)p^*\right)^T}{\left(p(1-p^*)\right)^T-\left((1-p)p^*\right)^T}(1-\gamma)^{-1}\right].$$

If conversely $\lambda \in I_1$, then

$$g(\lambda,b) = b(1+\lambda)+\lambda\mu+\left(\frac{1-p}{1-p^*}\right)^T(\lambda+(1-\gamma)^{-1})(v+b)^- - \left(\frac{p}{p^*}\right)^T\lambda(v+b)^+.$$

Therefore, $g(\lambda,\cdot)$ is a piecewise linear function with slope

$$\frac{\partial g(\lambda,b)}{\partial b} = \begin{cases} 1+\lambda-\left(\frac{1-p}{1-p^*}\right)^T(\lambda+(1-\gamma)^{-1}) &, \quad b<-v,\\ 1+\lambda-\left(\frac{p}{p^*}\right)^T\lambda &, \quad b>-v.\end{cases}$$

Hence, the infimum over all $b \in \mathbb{R}$ is finite – and then attained in $b^* = -v$ – if and only if for $\lambda \geq 0$

$$\begin{aligned}\inf_{b\in\mathbb{R}} g(\lambda,b) > -\infty &\Leftrightarrow 1+\lambda-\left(\frac{1-p}{1-p^*}\right)^T(\lambda+(1-\gamma)^{-1} \leq 0 \leq 1+\lambda-\left(\frac{p}{p^*}\right)^T\lambda \\ &\Leftrightarrow \lambda \leq \frac{(1-p)^T(1-\gamma)^{-1}-(1-p^*)^T}{(1-p^*)^T-(1-p)^T} \wedge \lambda \leq \frac{(p^*)^T}{p^T-(p^*)^T} \\ &\Leftrightarrow \lambda \leq C_\gamma(p,p^*).\end{aligned}$$

Note that $b^* = -v$ is the unique optimal point if and only if $\lambda < C_\gamma(p,p^*)$. We thoroughly treat the uniqueness below.

If $\gamma < \gamma_1(p,p^*)$ we have $C_\gamma(p,p^*) < 0$ and therefore $I_2 := [0, C_\gamma(p,p^*)] = \emptyset$. Consequently, we have for $\lambda \in \mathbb{R}_+ \setminus I_2$

$$\inf_{\pi\in F^T} L(\pi,\lambda) = \inf_{b\in\mathbb{R}} g(\lambda,b) = -\infty. \tag{2.13}$$

Now, let $\gamma \geq \gamma_1(p,p^*)$, so $I_2 \neq \emptyset$, and $\lambda \in I_2$. A simple calculation yields $I_2 \subset I_1$:

$$\begin{aligned}&\frac{(p^*)^T}{p^T-(p^*)^T} > \frac{\big((1-p)p^*\big)^T}{\big(p(1-p^*)\big)^T-\big((1-p)p^*\big)^T}(1-\gamma)^{-1} \\ \Leftrightarrow\ &\gamma < p^T\frac{(1-p^*)^T-(1-p)^T}{\big(p(1-p^*)\big)^T-\big((1-p)p^*\big)^T} = \gamma_2(p,p^*) \qquad (2.14) \\ \Leftrightarrow\ &\frac{(1-p)^T(1-\gamma)^{-1}-(1-p^*)^T}{(1-p^*)^T-(1-p)^T} < \frac{\big((1-p)p^*\big)^T}{\big(p(1-p^*)\big)^T-\big((1-p)p^*\big)^T}(1-\gamma)^{-1},\end{aligned}$$

hence it always holds

$$C_\gamma(p,p^*) \leq \frac{\big((1-p)p^*\big)^T}{\big(p(1-p^*)\big)^T-\big((1-p)p^*\big)^T}(1-\gamma)^{-1}.$$

The infimum becomes for $\gamma \geq \gamma_1(p,p^*)$ and $\lambda \in I_2$ with $b^* = -v$ and $J_0^{\lambda,-v}(v) = 0$

$$\begin{aligned}\inf_{\pi\in F^T} L(\pi,\lambda) &= \inf_{b\in\mathbb{R}} g(\lambda,b) = g(\lambda,b^*) = -v(1+\lambda)+\lambda\mu+J_0^{\lambda,-v}(v) \\ &= (\mu-v)\lambda - v.\end{aligned}$$

Together with (2.13) this yields the first part of the assertion.

In this situation we see from the calculation above that an optimal policy π^*, i. e. with $L(\pi^*,\lambda) = (\mu-v)\lambda-v$, is given by $\pi^* = \pi_{-v}$. If $\lambda < C_\gamma(p,p^*)$ the policy is also unique since $g(\lambda,\cdot)$ is strictly decreasing on $(-\infty,-v)$ and strictly increasing on $(-v,\infty)$. However, for $\lambda = C_\gamma(p,p^*)$ this is not the case. We have seen that the function $g(\lambda,\cdot)$ is constant on $(-v,\infty)$ for $\gamma \geq \gamma_2(p,p^*)$ and constant on $(-\infty,-v)$ for $\gamma \leq \gamma_2(p,p^*)$. Hence, additionally every π_b for $b > -v$ or $b < -v$ respectively is optimal if $\lambda = C_\gamma(p,p^*)$. This concludes the proof. □

Now, the solution of (RM) is based on the following standard result from convex optimization theory, which follows immediately from Theorem 5, Chapter 1, §1.1 in Ioffe and Tihomirov (1979), for example.

Let X be a linear space and $A \subset X$ a convex subset. Introduce the problem

$$\begin{aligned} &\inf_{x \in A} g_0(x) \\ &\text{s.t. } g_1(x) \leq 0, \end{aligned} \tag{CO}$$

for convex functions $g_0, g_1 : X \to \mathbb{R}$. Furthermore, define the Lagrange function $L : X \times \mathbb{R}_+ \to \mathbb{R}$ via

$$L(x, \lambda) := g_0(x) + \lambda \cdot g_1(x), \quad (x, \lambda) \in X \times \mathbb{R}_+$$

Then (CO) is equivalent to

$$\inf_{x \in A} \sup_{\lambda \geq 0} L(x, \lambda). \tag{PP}$$

The dual problem is

$$\sup_{\lambda \geq 0} \inf_{x \in A} L(x, \lambda), \tag{DP}$$

which gives in general only a lower bound for the primal problem.

Theorem 2.3. *Assume that the* Slater condition *is fulfilled, i. e. that there exists $x \in A$ such that $g_1(x) < 0$. Then the following are equivalent for $x^* \in A$:*

(i) x^* is an optimal solution of (PP).

(ii) There exists $\lambda^* \geq 0$ such that (x^*, λ^*) is an optimal solution of (DP) and fulfills the Kuhn–Tucker conditions

$$L(x^*, \lambda^*) = \inf_{x \in A} L(x, \lambda^*), \quad g_1(x^*) \leq 0, \ \lambda^* g_1(x^*) = 0.$$

Furthermore, if (i) or (ii) holds, the optimal values of (PP) and (DP) are equal.

Theorem 2.3 can be applied to (RM) since V_T^π is linear in π. It follows that the target function of (RM) is convex in π due to subadditivity of the Average Value at Risk. Furthermore, the constraint is a linear function in π and the Slater condition is obviously fulfilled, too. Now, the following result holds.

Theorem 2.4 ((RM) with shortselling). *Consider the function h_1 defined in* (2.5) *and let $v \in S$. Then*

$$h_1(v) = \begin{cases} (\mu - v)^+ \cdot C_\gamma(p, p^*) - v & , \quad \gamma \geq \gamma_1(p, p^*), \\ -\infty & , \quad \textit{otherwise}. \end{cases}$$

Furthermore, in the finite case an optimal policy is π_{b^*} defined in Theorem 2.2 with

$$b^* = \begin{cases} (\mu - v)^+ \dfrac{(p^*)^T}{p^T - (p^*)^T} - v \quad , & \text{if } \gamma \geq \gamma_2(p, p^*) \text{ and } p > p^* \\ & \text{or } \gamma \leq \gamma_2(p, p^*) \text{ and } p < p^* \end{cases} ,$$
$$ \begin{cases} -(\mu - v)^+ \dfrac{(1 - p^*)^T}{(1 - p^*)^T - (1 - p)^T} - v \quad , & \text{if } \gamma \leq \gamma_2(p, p^*) \text{ and } p > p^* \\ & \text{or } \gamma \geq \gamma_2(p, p^*) \text{ and } p < p^* \end{cases} .$$

If $p \neq p^*$, $\mu > v$, $\gamma > \gamma_1(p, p^*)$ and $\gamma \neq \gamma_2(p, p^*)$ the given optimal policy is unique.

Proof. If $p = p^*$ Lemma 2.1 implies that there are no admissible policies for $v < \mu$. Hence, $h_1(v) = \infty = (\mu - v) \cdot C_\gamma(p, p^*)$. Conversely, in the case $v \geq \mu$, all policies are admissible and the problem is equivalent to the one in Proposition 2.1. Hence, it holds with the convention $0 \cdot \infty = 0$

$$h_1(v) = -v = 0 \cdot \infty - v = (\mu - v)^+ \cdot C_\gamma(p, p^*) - v$$

and π_{b^*} with $b^* = -v$ is optimal.

The rest of the proof is given for $p > p^*$. If $\gamma < \gamma_1(p, p^*)$ assume that there exists a solution π^* of (RM). By Theorem 2.3 there exists $\lambda^* \geq 0$ such that $\inf_{\pi \in F^T} L(\pi, \lambda^*) = L(\pi^*, \lambda^*) > -\infty$. But this a contradiction to Lemma 2.4, where we have seen that $\inf_{\pi \in F^T} L(\pi, \lambda) = -\infty$ for all $\lambda \geq 0$. Hence, no optimal policy π^* exists and consequently $h_1(v) = -\infty$, since obviously there always exist admissible policies.

Now assume $\gamma \geq \gamma_1(p, p^*)$. We obtain for the dual problem with Lemma 2.4

$$\begin{aligned} h_2(v) &= \sup_{\lambda \geq 0} \inf_{\pi \in F^T} L(\pi, \lambda) = \sup_{0 \leq \lambda \leq C_\gamma(p, p^*)} \big\{ (\mu - v)\lambda - v \big\} \\ &= (\mu - v)^+ \cdot C_\gamma(p, p^*) - v, \end{aligned}$$

where $\lambda^* = C_\gamma(p, p^*)$ if $\mu > v$ and $\lambda^* = 0$ otherwise. Furthermore, the set of policies π such that $L(\pi, \lambda^*) = h_2(v)$ is given by Lemma 2.4. If we can show that then there is such an optimal π so that additionally, the pair (π, λ^*) also fulfills the Kuhn–Tucker–conditions, then Theorem 2.3 yields $h_1(v) = h_2(v)$ and therefore the assertion.

First, consider the simpler case $\mu \leq v = V_0$, hence $\lambda^* = 0$. If $\gamma > \gamma_1(p, p^*)$ we have $C_\gamma(p, p^*) > 0 = \lambda^*$. Lemma 2.4 yields that a corresponding optimal policy is π_b with $b = -v$. This is clearly admissible since $V_T^{\pi_{-v}} = v$ and therefore $\mathbb{E}\,[V_T^{\pi^*} \,|\, V_0 = v] = v \geq \mu$, hence the pair (π^*, λ^*) fulfills the Kuhn–Tucker–conditions. If $\gamma = \gamma_1(p, p^*)$ we have $C_\gamma(p, p^*) = 0 = \lambda^*$. Again by Lemma 2.4 we have that π_{-v} is optimal with $b = -v$, for example. We have seen that it is admissible and therefore have completed this part.

Now suppose $\mu > v$, hence $\lambda^* = C_\gamma(p, p^*)$ and the corresponding set of optimal policies is given by Lemma 2.4. First, we aim to find $b \geq -v$ with

$$\mathbb{E}\,[V_T^{\pi_b} \,|\, V_0 = v] = \mu. \tag{2.15}$$

We will even show that there exists a unique $b_1^* \geq -v$ such that this is fulfilled and that every π_b with $b \geq b_1^*$ is admissible for (RM).

Recall that $\pi_b = (f_0, \dots, f_{T-1})$ with

$$f_t(v') = \max\Big\{ -\frac{v'+b}{u-1}, -\frac{v'+b}{d-1}\Big\}, \quad v' \in S, t = 0, 1, \dots, T-1.$$

By Lemma 2.3 we obtain

$$\mathbb{E}\,[V_T^{\pi_b} \,|\, V_0 = v] = v + (v+b) \cdot \frac{p^T - (p^*)^T}{(p^*)^T}.$$

Setting $\mathbb{E}\,[V_T^{\pi_b} \,|\, V_0 = v] = \mu$ and solving for b shows that

$$b_1^* := (\mu - v) \cdot \frac{(p^*)^T}{p^T - (p^*)^T} - v > -v$$

yields the unique $\pi^* = \pi_{b_1^*}$ from the class $\{\pi_b \,|\, b \geq -v\}$ such that (2.15) holds. Furthermore, every π_b with $b \geq b_1^*$ is admissible for (RM). One can show similarly that there exists a unique $b \leq -v$ with (2.15) defined by

$$b_2^* := -(\mu - v) \cdot \frac{(1-p^*)^T}{(1-p^*)^T - (1-p)^T} - v < -v$$

such that every π_b with $b \leq b_2^*$ is admissible for (RM).

Now, in order to check that a pair (π^*, λ^*) with $L(\pi^*, \lambda^*) = \inf_{\pi \in F^T} L(\pi, \lambda^*)$ fulfills the Kuhn–Tucker conditions we have to discern two cases. If $\gamma = \gamma_1(p, p^*)$ we have $\lambda^* = 0$. In this case we only have to find an optimal π^* that is also admissible for (RM). By Lemma 2.4 and the considerations from above this is fulfilled for every π_b with $b \leq b_2^*$, i. e. these policies are also optimal for the primal problem (RM).

Otherwise, i. e. if $\gamma > \gamma_1(p, p^*)$, we have $\lambda^* > 0$ and π^* has to fulfill the constraint of (RM) even with equality in order to ensure $\lambda^*(\mu - \mathbb{E}\,[V_T^{\pi^*} \,|\, V_0 = v]) = 0$. Again, Lemma 2.4 and the considerations from above yield that for $\gamma > \gamma_2(p, p^*)$ this can only be fulfilled for $\pi_{b_1^*}$ and for $\gamma_1(p, p^*) < \gamma < \gamma_2(p, p^*)$ by $\pi_{b_2^*}$. If $\gamma = \gamma_2(p, p^*)$, both policies are optimal. □

Remark. 1. This is of course the same result that we obtained in an elementary way for the one period case.

2. It was also shown that in the case $\mu \leq v$, i. e. if we have an initial wealth that is already at least as high as the desired minimal expected terminal wealth μ, it is optimal to invest nothing in the asset and the solution is the same as in the unconstrained case.

Interpretation of the optimal policy The optimal policy is of the type π_b for some $b \in \mathbb{R}$. It can be observed that for $p > p^*$ the invested amount in the risky asset is always positive and for $p < p^*$ it is always negative. This seems quite natural since e. g. for $p > p^*$ we expect an upward tendency of the asset and otherwise a downward development.

Furthermore, the invested amount a_t at every time $t \in \{0, 1, \ldots, T-1\}$ is always a fraction of $V_t + b$, where the value of $-b$ could be regarded as a constant claim similar to the model in Runggaldier et al. (2002). This fraction depends on whether the term $V_t + b$ is positive or negative but on the other hand does not change over time. As described in Remark 4.2 in Runggaldier et al. (2002) the policy can be regarded as a hedging strategy in the CRRM for a certain claim.

2.3 Utility maximization

In this section, the complementary problem (UM) and a variation introduced below and denoted by (UM') are investigated. The aim is to maximize the expected wealth under risk constraints. First, we briefly consider the problem (UM) as described above. Not surprisingly, this can be treated in exactly the same way as (RM). Afterwards, we discern the two cases with and without shortselling and show how the expected final wealth can be maximized with and without certain intermediate constraints.

2.3.1 Comparison with the risk minimization

The problem (UM) with $u = \mathrm{id}_{\mathbb{R}}$, $\rho = \mathrm{AV@R}_\gamma$ for some $\gamma \in (0,1)$ has the following solution, which is again attained by a Lagrangian approach. Denote the optimal target value of (UM) with $h(v)$.

Theorem 2.5 ((UM) with shortselling). *Let $v \in S$. Then*

$$h(v) = \begin{cases} (R+v) \cdot (C_\gamma(p,p^*))^{-1} + v & , \quad R \geq -v, \ \gamma \geq \gamma_1(p,p^*), \\ -\infty & , \quad R < -v, \ \gamma \geq \gamma_1(p,p^*), \\ +\infty & , \quad \textit{otherwise}. \end{cases}$$

Proof. In the same way as in Subsection 2.2.2 we have equality of primal and dual problem and

$$h(v) = \inf_{\lambda \geq 0} \sup_{b \in \mathbb{R}} \left\{ b(1+\lambda) + \lambda R - \inf_{\pi \in F^T} \mathbb{E}\left[J_T(V_T^\pi - b) \,|\, V_0 = v \right] \right\}$$

where J_T is as in Theorem 2.2, this time with

$$\alpha_1 := 1 + \frac{\lambda}{1-\gamma} > 1 =: \alpha_2.$$

The rest follows analogously to the proofs of Lemma 2.4 and Theorem 2.4. Note that the intervals that occur in the proofs are

$$I_1 = \left[\frac{\left(p(1-p^*)\right)^T - \left((1-p)p^*\right)^T}{\left((1-p)p^*\right)^T}(1-\gamma), \infty\right)$$

and

$$I_2 = \left[(C_\gamma(p,p^*))^{-1}, \infty\right).$$

□

In the case $R < -v$ and $\gamma \geq \gamma_1(p,p^*)$, the optimal (here: maximal) value of the expectation is $-\infty$, implying that no admissible policies exist. This seems a little odd at first glance, but looking at the one-period model illustrates indeed why this is the case. Let $p > p^*$. If $T = 1$, we have with $\pi = (f_0) \in F^1$, $f_0(v) = a$, $v \in S$, for arbitrary $a \in A$

$$\begin{aligned} \mathrm{AV@R}_\gamma(V_1^\pi \mid V_0 = v) &= \mathrm{AV@R}_\gamma(v + a(Y_1 - 1)) \\ &= \begin{cases} -v + a(\mathrm{AV@R}_\gamma(Y_1) + 1) &, \quad a \geq 0 \\ -v - a(\mathrm{AV@R}_\gamma(-Y_1) - 1) &, \quad a < 0. \end{cases} \end{aligned}$$

By Lemma 2.2, we have $\mathrm{AV@R}_\gamma(-Y_1) - 1 > 0$ and in the case

$$\gamma \geq \gamma_1(p,p^*) = \frac{p - p^*}{1 - p^*}$$

also $\mathrm{AV@R}_\gamma(Y_1) + 1 \geq 0$. Consequently, it holds that

$$\mathrm{AV@R}_\gamma(V_1^\pi \mid V_0 = v) \geq -v > R, \quad \pi \in F^1.$$

We now want to give a simple result comparing the problems (UM) and (RM) and show that the optimal policies for both problems are identical. They can be regarded as equivalent by choosing the parameters μ and R appropriately. This relationship is quite standard for related mean–variance portfolio optimization problems.

Proposition 2.2. *Let $\gamma \geq \gamma_1(p,p^*)$ and consider the problem (UM) with $R \in \mathbb{R}$ and $v \in S$ such that $R \geq -v$. Let π^* be an optimal policy for the problem (RM) with*

$$\mu^* := (R + v) \cdot (C_\gamma(p,p^*))^{-1} + v.$$

Then π^* also is a solution of (UM). Conversely, taking $\mu \in \mathbb{R}$ with $\mu \geq v$, an optimal policy for (UM) with

$$R^* := (\mu - v) \cdot C_\gamma(p,p^*) - v$$

is a solution of (RM).

Proof. If π^* is an optimal policy for (RM) with μ^* as defined above, we have by definition $\mu^* \geq v$ and obtain by Theorem 2.4

$$\begin{aligned}\text{AV@R}_\gamma(V_T^{\pi^*} \mid V_0 = v) &= (\mu^* - v) \cdot C_\gamma(p, p^*) - v \\ &= ((R + v) \cdot (C_\gamma(p, p^*))^{-1} + v - v) \cdot C_\gamma(p, p^*) - v \\ &= R + v - v = R.\end{aligned}$$

Hence, π^* is admissible for (UM). But since π^* is also admissible for (RM), it holds

$$\mathbb{E}\,[V_T^{\pi^*} \mid V_0 = v] \geq \mu^* = (R + v) \cdot (C_\gamma(p, p^*))^{-1} + v, \tag{2.16}$$

where the right hand side is just the optimal value of the target function of problem (UM). Hence, π^* is even optimal for (UM) with equality in (2.16). The proof of the second part works analogously. □

2.3.2 The unconstrained case

For the sake of completeness, we give here the very simple result on mean maximization without constraints. Formally, set $R = \infty$. Furthermore, we assume for the rest of this chapter that $p \neq p^*$.

Proposition 2.3. *Let $v \in S$. Then it holds:*

(i) With shortselling, we have

$$\sup_{\pi \in F^T} \mathbb{E}\,[V_T^\pi \mid V_0 = v] = \infty.$$

(ii) If shortselling is not allowed, we have

$$\sup_{\pi \in F_{ns}^T} \mathbb{E}\,[V_T^\pi \mid V_0 = v] = \begin{cases} v \cdot (\mathbb{E}\, Y_1)^T & , \quad p > p^* \\ v & , \quad p < p^* \end{cases}$$

In the case $p > p^*$, the unique optimal policy $\pi = (f_0, \dots, f_{T-1}) \in F_{ns}^T$ is given by $f_t(v') = v'$, $v' \in S$, $t = 0, 1, \dots, T-1$, whereas $\pi = (0, \dots, 0) \in F_{ns}^T$ is optimal when $p < p^*$.

Proof. (i) Choose $\pi = (0, \dots, 0, f_{T-1}) \in F^T$ with $f_{T-1}(v) = a$, $v \in S$, for an arbitrary constant $a \in \mathbb{R}$. We obtain

$$\mathbb{E}\,[V_T^\pi \mid V_0 = v] = v + a\mathbb{E}\,[Y_T - 1].$$

Depending on $p > p^*$ or $p < p^*$ we see that (compare (2.3)) we can choose a arbitrarily large or small respectively, hence the assertion.

(ii) Consider the Markov decision model from Section 2.1 with $D(v) = [0, v]$, $v \in S$, and $J_T(v) = v$. We obtain

$$\begin{aligned} J'_{T-1}(v) &= \sup_{a \in D(v)} \mathbb{E}\left[J'_T(V_T(a_{T-1})) \mid V_{T-1} = v, a_{T-1} = a\right] \\ &= \sup_{a \in D(v)} \{v + a\mathbb{E}\left[Y_T - 1\right]\} \\ &= \begin{cases} v \cdot \mathbb{E}\, Y_T & , \quad p > p^* \\ v & , \quad p < p^*, \end{cases} \end{aligned}$$

where $a^* = f^*_{T-1}(v) = v \in D(v)$ for $p > p^*$ and $a^* = f^*_{T-1}(v) = 0 \in D(v)$ for $p < p^*$. Observing that $\mathbb{E}\, Y_T > 0$ iteratively yields the assertion using Theorem 2.1. □

Interpretation of the optimal policy Part (ii) yields the natural result that without a constraint and without shortselling, one should invest at time t as much as possible, here the whole capital V_t, in the asset if $p > p^*$, i. e. $\mathbb{E}\left[Y_1 - 1\right] > 0$, and everything in the risk free bond otherwise.

2.3.3 Intermediate constraints

Now we aim to solve the following modification of (UM):

$$\begin{aligned} &\sup_{\pi \in F'} \mathbb{E}\left[V_T^\pi \mid V_0 = v\right] \\ &\text{s.t. } \mathrm{AV@R}_\gamma(V_t^\pi \mid V_{t-1}^\pi) \leq R(V_{t-1}^\pi),\ t = 1, \dots, T, \end{aligned} \qquad \text{(UM')}$$

where the *risk function* $R : S \to \mathbb{R}$ is some $(\mathcal{S}, \mathcal{B})$–measurable mapping modelling the risk constraint. We will see immediately why this construction is chosen. Again, the set F' can mean either F^T_{ns} or F^T.

A simple choice would be to set $R(v) := R$, $v \in S$, for fixed R. However, the idea of the model (UM') is that based on the current state, in every trading period from $t-1$ to t it is ensured that the risk does not exceed a certain maximal level. But after some periods the current wealth could already be very high. Thus, the risk constraint R could become redundant if its level is not adapted over time by using the current wealth. Indeed, we will model the function R such that we can ensure that the risk of the *relative* gain or loss of wealth from one point of time to the next one is not too high. We believe that this is a more reasonable approach.

No shortselling

If shortselling is not allowed, the wealth process only attains non–negative values, i. e. $S = \mathbb{R}_+$. Furthermore, we have $F' = F^T_{\mathrm{ns}}$. In this case,

$$D(v) \subset [0, v], \quad v \in S.$$

The risk constraint might make a further restriction necessary. To follow the approach described above define

$$R(v) := v \cdot R - v, \quad v \in S,$$

for given $R \in \mathbb{R}$. Using translation invariance and homogeneity of Average Value-at-Risk, the optimization problem becomes

$$\begin{aligned} &\sup_{\pi \in F_{\mathrm{ns}}^T} \mathbb{E}\left[V_T^\pi \mid V_0 = v\right] \\ &\text{s.t. } \mathrm{AV@R}_\gamma\left(\frac{V_t^\pi - V_{t-1}^\pi}{V_{t-1}^\pi} \,\middle|\, V_{t-1}^\pi\right) \leq R,\, t = 1, \ldots, T. \end{aligned} \tag{UM'}$$

Obviously, (UM') fits into the Markov decision model from Section 2.1 by setting $J_T(v) = v$, $v \in S$. The restriction set becomes for $v \in S$ and $p > p^*$

$$\begin{aligned} D(v) &= \left\{a \in [0, v] \,\middle|\, \mathrm{AV@R}_\gamma\left(\frac{V_t(a) - V_{t-1}}{V_{t-1}} \,\middle|\, V_{t-1} = v\right) \leq R\right\} \\ &= \left\{a \in [0, v] \,\middle|\, \mathrm{AV@R}_\gamma\left(\frac{a(Y_t - 1)}{v}\right) \leq R\right\} \\ &= \left\{a \in [0, v] \,\middle|\, a \cdot \mathrm{AV@R}_\gamma(Y_1 - 1) \leq v \cdot R\right\} \\ &= \begin{cases} \left[0, v \cdot \min\{1, \delta_{\mathrm{u}}(\gamma, R)\}\right] & , \quad \gamma \geq \frac{p - p^*}{1 - p^*}, \\ \left[v \cdot \max\{0, \delta_{\mathrm{u}}(\gamma, R)\}, v\right] & , \quad \gamma < \frac{p - p^*}{1 - p^*}, \end{cases} \end{aligned}$$

where we used Lemma 2.2 and set

$$\delta_{\mathrm{u}}(\gamma, R) := \frac{R}{\mathrm{AV@R}_\gamma(Y_1 - 1)}, \quad \delta_{\mathrm{l}}(\gamma, R) := \frac{R}{\mathrm{AV@R}_\gamma(-Y_1 + 1)}.$$

If $\gamma \geq \frac{p-p^*}{1-p^*}$ and $R < 0$, we have $D(v) = \emptyset$, and if $\gamma < \frac{p-p^*}{1-p^*}$, $R \geq 0$ implies $D(v) = [0, v]$. We conclude that the structure of $D(v)$ depends strongly on the relationship of γ and R, which determines whether $\delta_{\mathrm{u}}(\gamma, R) > 1$ or $\delta_{\mathrm{u}}(\gamma, R) \leq 1$ respectively.

Analogously, we obtain for $p < p^*$

$$D(v) = \left[0, v \cdot \min\{1, \delta_{\mathrm{u}}(\gamma, R)\}\right],$$

where again $D(v) = \emptyset$ if $R < 0$. We are now ready to solve (UM').

Theorem 2.6 ((UM') without shortselling). *Let $v \in S$. Then the solution of (UM') is given by $J_0(v)$ which attains the following values.*

(i) Let $p > p^*$. If $\gamma \geq \frac{p-p^*}{1-p^*}$ we have

$$J_0(v) = \begin{cases} v \cdot \left(1 + \mathbb{E}\left[Y_1 - 1\right] \cdot \min\{1, \delta_u(\gamma, R)\}\right)^T & , \quad R \geq 0, \\ -\infty & , \quad R < 0. \end{cases}$$

In the finite case, the unique optimal policy is given by $\pi = (f_0, \dots, f_{T-1})$ with

$$f_t(v') = v' \cdot \min\{1, \delta_u(\gamma, R)\}, \quad v' \in S, t = 0, 1, \dots, T-1.$$

For $\gamma < \frac{p-p^*}{1-p^*}$ it holds

$$J_0(v) = \begin{cases} v \cdot (\mathbb{E}\, Y_1)^T & , \quad R \geq \mathrm{AV@R}_\gamma(Y_1 - 1), \\ -\infty & , \quad R < \mathrm{AV@R}_\gamma(Y_1 - 1)., \end{cases}$$

and the unique optimal policy is as in Proposition 2.3 (ii).

(ii) Let $p < p^*$. We then have

$$J_0(v) = \begin{cases} v & , \quad R \geq 0, \\ -\infty & , \quad R < 0. \end{cases}$$

For $R \geq 0$ the optimal policy is again as in Proposition 2.3 (ii).

Proof. (i) Let $p > p^*$ and $R < 0$. If $\gamma \geq \frac{p-p^*}{1-p^*}$ or if $\gamma < \frac{p-p^*}{1-p^*}$ and $\delta_{\mathrm{u}}(\gamma, R) > 1$, i. e. $R < \mathrm{AV@R}_\gamma(Y_1 - 1) < 0$, the restriction set becomes $D(v) = \emptyset$. Hence

$$J'_{T-1}(v) = \sup_{a \in D(v)} \{v + a\mathbb{E}\left[Y_T - 1\right]\} = -\infty$$

and therefore also $J_0(v) = J'_0(v) = -\infty$. If $\gamma < \frac{p-p^*}{1-p^*}$ and $\delta_{\mathrm{u}}(\gamma, R) \leq 1$ it holds

$$D(v) = \left[v \cdot \delta_{\mathrm{u}}(\gamma, R), v\right].$$

As in the proof of Proposition 2.3 (ii) we obtain $J_0(v) = v \cdot (\mathbb{E}\, Y_1)^T$.

Now, let $R \geq 0$. If $\gamma < \frac{p-p^*}{1-p^*}$ we have $D(v) = [0, v]$ and in the same way as above this yields $J_0(v) = v \cdot (\mathbb{E}\, Y_1)^T$. In the case $\gamma \geq \frac{p-p^*}{1-p^*}$ it holds

$$D(v) = \left[0, v \cdot \min\{1, \delta_{\mathrm{u}}(\gamma, R)\}\right].$$

We obtain

$$\begin{aligned} J'_{T-1}(v) &= \sup_{a \in D(v)} \mathbb{E}\left[J_T(V_T(a_{T-1})) \mid V_{T-1} = v, a_{T-1} = a\right] \\ &= \sup_{a \in D(v)} \{v + a \underbrace{\mathbb{E}\left[Y_T - 1\right]}_{>0}\} \\ &= v + \mathbb{E}\left[Y_T - 1\right] \cdot \sup_{a \in D(v)} a \qquad (2.17) \\ &= v + \mathbb{E}\left[Y_T - 1\right] \cdot v \cdot \min\{1, \delta_{\mathrm{u}}(\gamma, R)\} \\ &= v \cdot \left(1 + \mathbb{E}\left[Y_1 - 1\right] \cdot \min\{1, \delta_{\mathrm{u}}(\gamma, R)\}\right). \end{aligned}$$

This iteratively yields by Theorem 2.1

$$J_0(v) = J_0'(v) = v \cdot \big(1 + \mathbb{E}[Y_1 - 1] \cdot \min\{1, \delta_u(\gamma, R)\}\big)^T.$$

(ii) Let $p < p^*$ and $R < 0$. Then $D(v) = \emptyset$ and as above $J_0(v) = -\infty$. If $R \geq 0$ it holds

$$D(v) = \big[0, v \cdot \min\{1, \delta_u(\gamma, R)\}\big]$$

and the assertion follows as in the proof of Proposition 2.3 (ii). □

Remark. It can be observed from the proof that a further generalization is possible. Indeed, the parameters R and γ can be chosen time–dependent if we assume $R_t \geq 0$, $t = 1, \ldots, T$. The result would change for $\gamma_t \geq \frac{p-p^*}{1-p^*}$, $p > p^*$, where now

$$J_0(v) = v \cdot \prod_{t=1}^{T} \big(1 + \mathbb{E}[Y_t - 1] \cdot \min\{1, \delta_u(\gamma, R)\}\big).$$

This is possible since Theorem 2.1 is also valid if the restriction set is time–dependent (compare Hernández-Lerma and Lasserre (1996), Section 3.4).

Interpretation of the optimal policy First, let $p > p^*$, i. e. we have an upward tendency in the asset, and let the safety level γ be sufficiently large. Then there are either no admissible policies (if we only allow for negative risk at every time step) or it is optimal to invest a fraction of the current wealth into the risky asset. This fraction can be smaller than the current wealth (namely if $\delta_u(\gamma, R) < 1$), hence we also invest a positive amount in the risk–free bond, in contrast to the unconstrained case. This is of course due to the constraint which forces the investor to be more cautious in this situation. However, if the constant R is chosen too large, the constraints become redundant and the result is equivalent to the one from Proposition 2.3 (ii). This argument also applies for small safety levels.

If $p < p^*$ there are again no admissible policies if we are too careful when choosing $R < 0$. For $R \geq 0$ we again have to invest everything in the risk–free bond, i. e. our behaviour does not change compared to the unconstrained case.

Compared to the situation with only a final constraint we see that the structure of the optimal policy is much simpler and more natural to interpret.

With shortselling

If shortselling is allowed the proofs become more complicated. But the structure of the optimal value of the target function is very simple. Now, the current wealth can also attain negative values. Hence, we consider (UM') with $S = \mathbb{R}$ and $F' = F^T$ and define the risk function R through

$$R(v) = |v| \cdot R - v, \quad v \in S,$$

for given $R \in \mathbb{R}$ such that we have to solve

$$\begin{aligned} &\sup_{\pi \in F_{\text{ns}}^T} \mathbb{E}\left[V_T^\pi \mid V_0 = v\right] \\ &\text{s.t. } \text{AV@R}_\gamma\left(\frac{V_t^\pi - V_{t-1}^\pi}{|V_{t-1}^\pi|} \,\middle|\, V_{t-1}^\pi\right) \le R,\ t = 1, \dots, T. \end{aligned} \tag{UM'}$$

Obviously, (UM') fits into the Markov decision model from Section 2.1 by setting $J_T(v) = v$, $v \in S$. The restriction set becomes for $v \in S$, $p > p^*$ and $\gamma \ge \frac{p-p^*}{1-p^*}$, again with Lemma 2.2,

$$\begin{aligned} &D(v) \\ &= \left\{a \in \mathbb{R} \,\middle|\, \text{AV@R}_\gamma\left(\frac{V_t(a) - V_{t-1}}{|V_{t-1}|} \,\middle|\, V_{t-1} = v\right) \le R\right\} \\ &= \left\{a \in \mathbb{R} \,\middle|\, \text{AV@R}_\gamma\left(\frac{a(Y_1 - 1)}{|v|}\right) \le R\right\} \\ &= \left\{a < 0 \,\middle|\, a \ge -|v| \cdot \frac{R}{\text{AV@R}_\gamma(-Y_1 + 1)}\right\} \cup \left\{a \ge 0 \,\middle|\, a \le |v| \cdot \frac{R}{\text{AV@R}_\gamma(Y_1 - 1)}\right\} \\ &= \big[-|v| \cdot \delta_{\text{l}}(\gamma, R)\,,\, |v| \cdot \delta_{\text{u}}(\gamma, R)\big]. \end{aligned}$$

Note that for $R < 0$ the restriction set becomes the empty set and therefore $J_0 = -\infty$. If $\gamma < \frac{p-p^*}{1-p^*}$ it holds $\text{AV@R}_\gamma(Y_1 - 1) < 0$ and we obtain in a similar way as above

$$\begin{aligned} &D(v) \\ &= \left\{a < 0 \,\middle|\, a \ge -|v| \cdot \frac{R}{\text{AV@R}_\gamma(-Y_1 + 1)}\right\} \cup \left\{a \ge 0 \,\middle|\, a \ge |v| \cdot \frac{R}{\text{AV@R}_\gamma(Y_1 - 1)}\right\} \\ &= \big[-|v| \cdot \delta_{\text{l}}(\gamma, R)\,,\, 0\big] \cup \big[\max\{0, |v| \cdot \delta_{\text{u}}(\gamma, R)\}\,,\, \infty\big). \end{aligned}$$

Hence, $a \in D(v)$ can be chosen arbitrarily large and by (2.17) we have $J'_{T-1} \equiv \infty$ and therefore also $J_0 \equiv \infty$. We conclude that if $p > p^*$ we only have to consider the case $\gamma \ge \frac{p-p^*}{1-p^*}$ and $R \ge 0$.

On the other hand, if $p < p^*$ we obtain $J_0 \equiv \infty$ if $\gamma < \frac{p^*-p}{p^*}$ and $J_0 \equiv -\infty$ if $\gamma \ge \frac{p^*-p}{p^*}$ and $R < 0$. Only the case $\gamma \ge \frac{p^*-p}{p^*}$ and $R \ge 0$ yields a non trivial solution with the same restriction set as above

$$D(v) = \big[-|v| \cdot \delta_{\text{l}}(\gamma, R)\,,\, |v| \cdot \delta_{\text{u}}(\gamma, R)\big].$$

We will now show that with $J_T(v) = v$, $v \in S$, the value functions introduced in Section 2.1 have the following structure

$$J'_t(v) = \alpha_+^{(t)} v^+ - \alpha_-^{(t)} v^-, \quad v \in S,\ t = 0, 1, \dots, T-1, \tag{2.18}$$

where the factors fulfill $\alpha_+^{(t)} > 0$ and $\alpha_+^{(t)} > \alpha_-^{(t)}$, but possibly $\alpha_-^{(t)} < 0$. Again by Theorem 2.1, the function J_0' provides the solution of (UM'). As usual, we thoroughly investigate the case $p > p^*$ and only state the corresponding result if $p < p^*$. We will now define the factors from (2.18) and a corresponding policy $\pi^* = (f_0^*, \dots, f_{T-1}^*)$. It is shown in a lemma below that the factors have the desired properties and in Theorem 2.7 that π^* is indeed optimal. First, let us define the following one-step policies for $v \in S$:

$$f^{(1)}(v) := |v| \cdot \delta_{\mathrm{u}}(\gamma, R) \in D(v),$$
$$f^{(2)}(v) := -|v| \cdot \delta_{\mathrm{l}}(\gamma, R) \in D(v)$$

Algorithm. *First, let $p > p^*$, $\gamma \geq \frac{p-p^*}{1-p^*}$ and $R \geq 0$. As initial values we have the one-step policy $f_{T-1}^* := f^{(1)}$ and*

$$\alpha_+^{(T-1)} := 1 + \mathbb{E}[Y_1 - 1] \cdot \delta_u(\gamma, R), \quad \alpha_-^{(T-1)} := 1 - \mathbb{E}[Y_1 - 1] \cdot \delta_u(\gamma, R).$$

Assume now we are given $\alpha_+^{(t)}$ and $\alpha_-^{(t)}$ with $\alpha_+^{(t)} > 0$ and $\alpha_+^{(t)} > \alpha_-^{(t)}$ for some $t \in \{1, \dots, T-1\}$. We want to define $\alpha_+^{(t-1)}$, $\alpha_-^{(t-1)}$ and a one-step policy f_{t-1}^*.

There are four possible values for the factor $\alpha_+^{(t-1)}$. First, define

$$\alpha_{+,11} := \alpha_+^{(t)} + \delta_u(\gamma, R) \cdot (u-d) \cdot \alpha_+^{(t)} \cdot \big(p(1-p^*) - (1-p)p^*\big),$$
$$\alpha_{+,12} := p\alpha_+^{(t)} + (1-p)\alpha_-^{(t)} + \delta_u(\gamma, R) \cdot (u-d) \cdot \big(\alpha_+^{(t)} p(1-p^*) - \alpha_-^{(t)}(1-p)p^*\big),$$
$$\alpha_{+,21} := \alpha_+^{(t)} + \delta_l(\gamma, R) \cdot (u-d) \cdot \alpha_+^{(t)} \cdot \big((1-p)p^* - p(1-p^*)\big),$$
$$\alpha_{+,22} := (1-p)\alpha_+^{(t)} + p\alpha_-^{(t)} + \delta_l(\gamma, R) \cdot (u-d) \cdot \big(\alpha_+^{(t)}(1-p)p^* - \alpha_-^{(t)} p(1-p^*)\big),$$

and then

$$\alpha_{+,1}^{(t-1)} := \begin{cases} \alpha_{+,11} & , \quad \delta_u(\gamma, R) \leq \dfrac{1}{1-d}, \\ \alpha_{+,12} & , \quad \delta_u(\gamma, R) > \dfrac{1}{1-d}, \end{cases}$$
$$\alpha_{+,2}^{(t-1)} := \begin{cases} \alpha_{+,21} & , \quad \delta_l(\gamma, R) \leq \dfrac{1}{u-1}, \\ \alpha_{+,22} & , \quad \delta_l(\gamma, R) > \dfrac{1}{u-1}. \end{cases}$$

There are also four possibilities for the factor $\alpha_-^{(t-1)}$. First, define

$$\alpha_{-,11} := \alpha_-^{(t)} - \delta_u(\gamma, R) \cdot (u-d) \cdot \alpha_-^{(t)} \cdot \big(p(1-p^*) - (1-p)p^*\big),$$
$$\alpha_{-,12} := p\alpha_+^{(t)} + (1-p)\alpha_-^{(t)} - \delta_u(\gamma, R) \cdot (u-d) \cdot \big(\alpha_+^{(t)} p(1-p^*) - \alpha_-^{(t)}(1-p)p^*\big),$$
$$\alpha_{-,21} := \alpha_-^{(t)} - \delta_l(\gamma, R) \cdot (u-d) \cdot \alpha_-^{(t)} \cdot \big((1-p)p^* - p(1-p^*)\big),$$
$$\alpha_{-,22} := (1-p)\alpha_+^{(t)} + p\alpha_-^{(t)} - \delta_l(\gamma, R) \cdot (u-d) \cdot \big(\alpha_+^{(t)}(1-p)p^* - \alpha_-^{(t)} p(1-p^*)\big).$$

and then

$$\alpha_{-,1}^{(t-1)} := \begin{cases} \alpha_{-,11} & , \quad \delta_u(\gamma, R) \leq \dfrac{1}{u-1}, \\ \alpha_{-,12} & , \quad \delta_u(\gamma, R) > \dfrac{1}{u-1}, \end{cases}$$

$$\alpha_{-,2}^{(t-1)} := \begin{cases} \alpha_{-,21} & , \quad \delta_l(\gamma, R) \leq \dfrac{1}{1-d}, \\ \alpha_{-,22} & , \quad \delta_l(\gamma, R) > \dfrac{1}{1-d}. \end{cases}$$

Now, we define $\alpha_+^{(t-1)}$ *and* $f_{t-1}^* : S \to A$ *such that* $f_{t-1}^*(v) \in D(v)$ *for all* $v \in S$ *with* $v \geq 0$ *as follows (see below for* $\alpha_-^{(t-1)}$ *and the case* $v < 0$*). In a first step assume* $\frac{p^*}{p}\alpha_+^{(t)} \leq \frac{1-p^*}{1-p}\alpha_-^{(t)}$. *Then* $f_{t-1}^* := f^{(1)}$ *and* $\alpha_+^{(t-1)} := \alpha_{+,1}^{(t-1)}$.

If on the other hand $\frac{p^*}{p}\alpha_+^{(t)} > \frac{1-p^*}{1-p}\alpha_-^{(t)}$, *two cases have to be discerned:*

(i) If $\delta_l(\gamma, R) \leq \frac{1}{u-1}$, *then again* $f_{t-1}^* := f^{(1)}$ *and* $\alpha_+^{(t-1)} := \alpha_{+,1}^{(t-1)}$.

(ii) If $\delta_l(\gamma, R) > \frac{1}{u-1}$ *set* $\alpha_+^{(t-1)} := \max\{\alpha_{+,1}^{(t-1)}, \alpha_{+,2}^{(t-1)}\}$ *and* $f_{t-1}^* := f^{(1)}$ *if it holds* $\alpha_+^{(t-1)} = \alpha_{+,1}^{(t-1)}$ *and* $f_{t-1}^* := f^{(2)}$ *otherwise.*

Now, we want to define $\alpha_-^{(t-1)}$ *and* $f_{t-1}^* : S \to A$ *such that* $f_{t-1}^*(v) \in D(v)$ *for all* $v \in S$ *with* $v < 0$ *as follows. In a first step assume again* $\frac{p^*}{p}\alpha_+^{(t)} \leq \frac{1-p^*}{1-p}\alpha_-^{(t)}$, *implying* $\alpha_-^{(t)} > 0$. *Then set* $f_{t-1}^* := f^{(1)}$ *and* $\alpha_-^{(t-1)} := \alpha_{-,1}^{(t-1)}$.

If on the other hand $\frac{p^*}{p}\alpha_+^{(t)} > \frac{1-p^*}{1-p}\alpha_-^{(t)}$, *again several cases have to be discerned:*

(a) $\underline{\alpha_-^{(t)} \leq 0 :}$ *Then there are two cases.*

  (i) If $\delta_u(\gamma, R) \leq \frac{1}{u-1}$ *then* $f_{t-1}^* := f^{(2)}$ *and* $\alpha_-^{(t-1)} := \alpha_{-,2}^{(t-1)}$.

  (ii) If $\delta_u(\gamma, R) > \frac{1}{u-1}$ *we set* $\alpha_-^{(t-1)} := \max\{\alpha_{-,1}^{(t-1)}, \alpha_{-,2}^{(t-1)}\}$ *and* $f_{t-1}^* := f^{(1)}$ *if* $\alpha_-^{(t-1)} = \alpha_{-,1}^{(t-1)}$ *and* $f_{t-1}^* := f^{(2)}$ *otherwise.*

(b) $\underline{\alpha_-^{(t)} > 0 :}$ *Then there are again two cases.*

  (i) If $\delta_l(\gamma, R) \leq \frac{1}{1-d}$ *then* $f_{t-1}^* := f^{(1)}$ *and* $\alpha_-^{(t-1)} := \alpha_{-,1}^{(t-1)}$.

  (ii) If $\delta_l(\gamma, R) > \frac{1}{1-d}$, *then* $\alpha_-^{(t-1)} := \max\{\alpha_{-,1}^{(t-1)}, \alpha_{-,2}^{(t-1)}\}$. *and* $f_{t-1}^* := f^{(1)}$ *if* $\alpha_-^{(t-1)} = \alpha_{-,1}^{(t-1)}$ *and* $f_{t-1}^* := f^{(2)}$ *otherwise.*

Now, let $p < p^*$, $\gamma \geq \frac{p^*-p}{p^*}$ *and* $R \geq 0$. *As initial values we have the one-step policy* $f_{T-1}^* := f^{(2)}$ *and*

$$\alpha_+^{(T-1)} := 1 - \mathbb{E}\left[Y_1 - 1\right] \cdot \delta_l(\gamma, R), \ \alpha_-^{(T-1)} := 1 + \mathbb{E}\left[Y_1 - 1\right] \cdot \delta_l(\gamma, R).$$

Assume now we are given $\alpha_+^{(t)}$ and $\alpha_-^{(t)}$ with $\alpha_+^{(t)} > 0$ and $\alpha_+^{(t)} > \alpha_-^{(t)}$ for some $t \in \{1, \ldots, T-1\}$. We define $\alpha_+^{(t-1)}$ and $f_{t-1}^* : S \to A$ such that $f_{t-1}^*(v) \in D(v)$ for all $v \in S$ with $v \geq 0$ as follows (see below for the case $v < 0$). In a first step assume $\frac{p}{p^*}\alpha_+^{(t)} \leq \frac{1-p}{1-p^*}\alpha_-^{(t)}$. Then $f_{t-1}^* := f^{(2)}$ and $\alpha_+^{(t-1)} := \alpha_{+,2}^{(t-1)}$.

If on the other hand $\frac{p}{p^*}\alpha_+^{(t)} > \frac{1-p}{1-p^*}\alpha_-^{(t)}$, two cases have to be discerned:

(i) If $\delta_u(\gamma, R) \leq \frac{1}{1-d}$, then again $f_{t-1}^* := f^{(2)}$ and $\alpha_+^{(t-1)} := \alpha_{+,2}^{(t-1)}$.

(ii) If $\delta_u(\gamma, R) > \frac{1}{1-d}$ set $\alpha_+^{(t-1)} := \max\{\alpha_{+,1}^{(t-1)}, \alpha_{+,2}^{(t-1)}\}$ and $f_{t-1}^* := f^{(1)}$ if it holds $\alpha_+^{(t-1)} = \alpha_{+,1}^{(t-1)}$ and $f_{t-1}^* := f^{(2)}$ otherwise.

Now, we want to define $\alpha_-^{(t-1)}$ and $f_{t-1}^* : S \to A$ such that $f_{t-1}^*(v) \in D(v)$ for all $v \in S$ with $v < 0$ as follows. In a first step assume again $\frac{p}{p^*}\alpha_+^{(t)} \leq \frac{1-p}{1-p^*}\alpha_-^{(t)}$, implying $\alpha_-^{(t)} > 0$. Then set $f_{t-1}^* := f^{(2)}$ and $\alpha_-^{(t-1)} := \alpha_{-,2}^{(t-1)}$.

If on the other hand $\frac{p}{p^*}\alpha_+^{(t)} > \frac{1-p}{1-p^*}\alpha_-^{(t)}$, again several cases have to be discerned:

(a) $\underline{\alpha_-^{(t)} \leq 0:}$ Then there are two cases.

- *(i) If $\delta_l(\gamma, R) \leq \frac{1}{1-d}$ then $f_{t-1}^* := f^{(1)}$ and $\alpha_-^{(t-1)} := \alpha_{-,1}^{(t-1)}$.*
- *(ii) If $\delta_l(\gamma, R) > \frac{1}{1-d}$ we set $\alpha_-^{(t-1)} := \max\{\alpha_{-,1}^{(t-1)}, \alpha_{-,2}^{(t-1)}\}$ and $f_{t-1}^* := f^{(1)}$ if $\alpha_-^{(t-1)} = \alpha_{-,1}^{(t-1)}$ and $f_{t-1}^* := f^{(2)}$ otherwise.*

(b) $\underline{\alpha_-^{(t)} > 0:}$ Then there are again two cases.

- *(i) If $\delta_u(\gamma, R) \leq \frac{1}{u-1}$ then $f_{t-1}^* := f^{(2)}$ and $\alpha_-^{(t-1)} := \alpha_{-,2}^{(t-1)}$.*
- *(ii) If $\delta_u(\gamma, R) > \frac{1}{u-1}$, then $\alpha_-^{(t-1)} := \max\{\alpha_{-,1}^{(t-1)}, \alpha_{-,2}^{(t-1)}\}$. and $f_{t-1}^* := f^{(1)}$ if $\alpha_-^{(t-1)} = \alpha_{-,1}^{(t-1)}$ and $f_{t-1}^* := f^{(2)}$ otherwise.*

We have introduced these factors and the optimal policy before proving the corresponding theorem below because the inductive proof only works if we can verify the desired properties. This is done in the following lemma.

Lemma 2.5. *For all $t \in \{0, 1, \ldots, T-1\}$ it holds*

$$\alpha_+^{(t)} > 0 \text{ and } \alpha_+^{(t)} > \alpha_-^{(t)}.$$

Proof. We only consider the case $p > p^*$. The proof is by backward induction on t where the case $t = T-1$ is obvious.

Now assume that the assertion holds for some $t \in \{1, \ldots, T-1\}$ and define $\alpha_+^{(t-1)}$ and $\alpha_-^{(t-1)}$ by the previous algorithm. First we want to show that $\alpha_+^{(t-1)} > 0$, where $\alpha_+^{(t-1)} \in \{\alpha_{+,1}^{(t-1)}, \alpha_{+,2}^{(t-1)}\}$. Obviously, in the case $\delta_u(\gamma, R) \leq \frac{1}{1-d}$ we have

$\alpha_{+,1}^{(t-1)} > 0$ by induction hypothesis and $p > p^*$. Moreover, note it always holds $\alpha_+^{(t)} p(1-p^*) - \alpha_-^{(t)}(1-p)p^* > 0$ since $p > p^*$. So if conversely $\delta_{\mathrm{u}}(\gamma, R) > \frac{1}{1-d}$ we obtain

$$\begin{aligned}
\alpha_+^{(t-1)} &= p\alpha_+^{(t)} + (1-p)\alpha_-^{(t)} + \frac{R(u-d)}{\mathrm{AV@R}_\gamma(Y_1 - 1)} \cdot \left(\alpha_+^{(t)} p(1-p^*) - \alpha_-^{(t)}(1-p)p^*\right) \\
&> p\alpha_+^{(t)} + (1-p)\alpha_-^{(t)} + \frac{u-d}{1-d} \cdot \left(\alpha_+^{(t)} p(1-p^*) - \alpha_-^{(t)}(1-p)p^*\right) \\
&= p\alpha_+^{(t)} + (1-p)\alpha_-^{(t)} + \frac{1}{p^*} \cdot \left(\alpha_+^{(t)} p(1-p^*) - \alpha_-^{(t)}(1-p)p^*\right) \\
&= p\alpha_+^{(t)} + \alpha_+^{(t)} \frac{p}{p^*}(1-p^*) > 0.
\end{aligned}$$

Finally, $\alpha_+^{(t-1)} = \alpha_{+,2}^{(t-1)}$ can only occur if $\alpha_{+,2}^{(t-1)} > \alpha_{+,1}^{(t-1)} > 0$. Hence, the first part of the assertion follows.

Now, we want to show $\alpha_+^{(t)} > \alpha_-^{(t)}$. First, let $\frac{p^*}{p}\alpha_+^{(t)} \leq \frac{1-p^*}{1-p}\alpha_-^{(t)}$. We then have $\alpha_+^{(t-1)} \in \{\alpha_{+,11}, \alpha_{+,12}\}$ and $\alpha_-^{(t-1)} \in \{\alpha_{-,11}, \alpha_{-,12}\}$, where pairwise comparison of the values obviously yields the assertion.

Now, let $\frac{p^*}{p}\alpha_+^{(t)} > \frac{1-p^*}{1-p}\alpha_-^{(t)}$. First note that then we always have

$$\begin{aligned}
&\alpha_{-,11} < \alpha_{+,11}, \alpha_{+,12}, \alpha_{+,2}^{(t-1)}, \\
&\alpha_{-,12} < \alpha_{+,11}, \alpha_{+,12}, \\
&\alpha_{-,21} < \alpha_{+,11}, \\
&\alpha_{-,22} < \alpha_{+,11}, \alpha_{+,2}^{(t-1)},
\end{aligned}$$

such that we can skip these relations in the investigations below.

Let us show $\alpha_{-,12} < \alpha_{+,2}^{(t-1)}$. The case $\alpha_+^{(t-1)} = \alpha_{+,2}$ can only occur if the relation $\delta_{\mathrm{l}}(\gamma, R) > \frac{1}{u-1}$ holds and $\alpha_-^{(t-1)} = \alpha_{-,12}$ only if $\delta_{\mathrm{u}}(\gamma, R) > \frac{1}{u-1}$. We obtain in this situation

$$\begin{aligned}
\alpha_{+,2}^{(t-1)} &= (1-p)\alpha_+^{(t)} + p\alpha_-^{(t)} + \frac{R(u-d)}{\mathrm{AV@R}_\gamma(-Y_1 + 1)} \cdot \left(\alpha_+^{(t)}(1-p)p^* - \alpha_-^{(t)} p(1-p^*)\right) \\
&> (1-p)\alpha_+^{(t)} + p\alpha_-^{(t)} + \frac{u-d}{u-1} \cdot \left(\alpha_+^{(t)}(1-p)p^* - \alpha_-^{(t)} p(1-p^*)\right) \\
&= (1-p)\alpha_+^{(t)} + p^* \frac{1-p}{1-p^*}\alpha_+^{(t)} > 0
\end{aligned}$$

and

$$\begin{aligned}
\alpha_{-,12} &= p\alpha_+^{(t)} + (1-p)\alpha_-^{(t)} - \frac{R(u-d)}{\mathrm{AV@R}_\gamma(Y_1 - 1)} \cdot \left(\alpha_+^{(t)} p(1-p^*) - \alpha_-^{(t)}(1-p)p^*\right) \\
&< p\alpha_+^{(t)} + (1-p)\alpha_-^{(t)} - \frac{u-d}{u-1} \cdot \left(\alpha_+^{(t)} p(1-p^*) - \alpha_-^{(t)}(1-p)p^*\right) \\
&= (1-p)\alpha_-^{(t)} + p^* \frac{1-p}{1-p^*}\alpha_-^{(t)} \leq 0,
\end{aligned}$$

hence the assertion. The fact that $\alpha_{-,22} < \alpha_{+,12}$ is shown similarly.

Let us finally show that $\alpha_{-,21} < \alpha_{+,12}, \alpha_{+,2}^{(t-1)}$. By construction, $\alpha_{-}^{(t-1)} = \alpha_{-,21}$ can only occur if $\alpha_{-}^{(t)} \leq 0$. But this already yields

$$\alpha_{-,21} = \alpha_{-}^{(t)} + \underbrace{\delta_{\mathrm{l}}(\gamma, R)}_{\geq 0} \cdot (u-d) \cdot \alpha_{-}^{(t)} \cdot \big(p(1-p^*) - (1-p)p^*\big) \leq \alpha_{-}^{(t)} < \alpha_{+,12}, \alpha_{+,2}^{(t-1)},$$

thus completing the proof. □

We are now ready to state and proof the final important result of this chapter.

Theorem 2.7 ((UM') with shortselling). *Let $v \in S$ and $R \geq 0$. Furthermore, assume that $\gamma \geq \frac{p-p^*}{1-p^*}$ if $p > p^*$ and $\gamma \geq \frac{p^*-p}{p^*}$ if $p < p^*$. Then the solution of (UM') is given by*

$$J_0(v) = \alpha_{+}^{(0)} v^+ - \alpha_{-}^{(0)} v^-,$$

where $\alpha_{+}^{(0)}$ and $\alpha_{-}^{(0)}$ are defined by the algorithm above which also provides an optimal policy.

Proof. Let $p > p^*$. We claim that

$$J_t'(v) = \alpha_{+}^{(t)} v^+ - \alpha_{-}^{(t)} v^-, \quad v \in S,\, t = 0, 1, \ldots, T-1. \tag{2.19}$$

By Theorem 2.1, the assertion will follow. The proof is by backward induction on t. Consider first $t = T-1$. We obtain again

$$\begin{aligned}
J_{T-1}'(v) &= \sup_{a \in D(v)} \mathbb{E}\left[J_T(V_T(a_{T-1})) \,|\, V_{T-1} = v, a_{T-1} = a\right] = \sup_{a \in D(v)} \left\{v + a\mathbb{E}\left[Y_T - 1\right]\right\} \\
&= v + \mathbb{E}\left[Y_T - 1\right] \sup_{a \in D(v)} a = v + \mathbb{E}\left[Y_T - 1\right] \cdot |v| \cdot \delta_{\mathrm{u}}(\gamma, R) \\
&= \Big(1 + \mathbb{E}\left[Y_T - 1\right] \cdot \delta_{\mathrm{u}}(\gamma, R)\Big) v^+ - \Big(1 - \mathbb{E}\left[Y_T - 1\right] \cdot \delta_{\mathrm{u}}(\gamma, R)\Big) v^- \\
&= \alpha_{+}^{(T-1)} v^+ - \alpha_{-}^{(T-1)} v^-.
\end{aligned}$$

One also sees that a maximizer is given by f_{T-1}^*. Hence, (2.19) holds for $t = T-1$.

Assume now that (2.19) is true for some $t \in \{1, \ldots, T-1\}$ and let $v \in S$. We have to solve

$$\begin{aligned}
J_{t-1}'(v) &= \sup_{a \in D(v)} \mathbb{E}\left[J_t'(V_t(a_{t-1})) \,\middle|\, V_{t-1} = v,\, a_{t-1} = a\right] \\
&= \sup_{a \in D(v)} \Big\{p\big(\alpha_{+}^{(t)}(v + a(u-1))^+ - \alpha_{-}^{(t)}(v + a(u-1))^-\big) \\
&\qquad + (1-p)\big(\alpha_{+}^{(t)}(v + a(d-1))^+ - \alpha_{-}^{(t)}(v + a(d-1))^-\big)\Big\}.
\end{aligned}$$

Let us first consider the case $v \geq 0$ and recall that the restriction set is given by

$$D(v) = \big[-v \cdot \delta_{\mathrm{l}}(\gamma, R)\,,\, v \cdot \delta_{\mathrm{u}}(\gamma, R)\big].$$

To determine the slope of the piecewise linear function that we have to maximize define

$$\psi^{(1)} := -\frac{v}{u-1} \leq 0 \leq -\frac{v}{d-1} =: \psi^{(2)}.$$

Similar to the proof of Theorem 2.2, the slope on $(\psi^{(2)}, \infty)$ is given by

$$\begin{aligned} & p\alpha_+^{(t)}(u-1) + (1-p)\alpha_-^{(t)}(d-1) > 0 \\ \Leftrightarrow \quad & \frac{p}{p^*}\alpha_+^{(t)} > \frac{1-p}{1-p^*}\alpha_-^{(t)}, \end{aligned}$$

which holds because $p > p^*$ and $\alpha_+^{(t)} > \alpha_-^{(t)}$ by Lemma 2.5. Hence, the function is increasing on this interval. This is also the case on $[\psi^{(1)}, \psi^{(2)}]$ because for the slope it holds

$$p\alpha_+^{(t)}(u-1) + (1-p)\alpha_+^{(t)}(d-1) > 0 \;\Leftrightarrow\; \alpha_+^{(t)}(p - p^*) > 0,$$

which is fulfilled since $\alpha_+^{(t)} > 0$ by Lemma 2.5. Finally, it follows for the slope on $(-\infty, \psi^{(1)})$

$$\begin{aligned} & p\alpha_-^{(t)}(u-1) + (1-p)\alpha_+^{(t)}(d-1) \geq 0 \\ \Leftrightarrow \quad & \frac{p^*}{p}\alpha_+^{(t)} \leq \frac{1-p^*}{1-p}\alpha_-^{(t)}, \end{aligned}$$

which can sometimes be fulfilled. If this is the case, the function is non-decreasing on $\mathbb{R}$ and consequently attains its maximum on $D(v)$ for its maximal admissible value. Hence, $f_{t-1,1}(v)$ is indeed optimal for a. The value of J'_{t-1} depends on whether $f_{t-1,1}(v) \leq \psi^{(2)}$ or $f_{t-1,1}(v) > \psi^{(2)}$. We have

$$f_{t-1,1}(v) \leq \psi^{(2)} \;\Leftrightarrow\; \delta_{\mathrm{u}}(\gamma, R) \leq \frac{1}{1-d} \tag{2.20}$$

If this is fulfilled, we have $v + f_{t-1,1}(v)(u-1) \geq 0$ and $v + f_{t-1,1}(v)(d-1) \geq 0$. Hence,

$$\begin{aligned} J'_{t-1}(v) &= p\alpha_+^{(t)}(v + f_{t-1,1}(v)(u-1)) + (1-p)\alpha_+^{(t)}(v + f_{t-1,1}(v)(d-1)) \\ &= \big(\alpha_+^{(t)} + \delta_{\mathrm{u}}(\gamma, R)\alpha_+^{(t)}(p(u-1) - (1-p)(1-d))\big) \cdot v \\ &= \alpha_{+,11} \cdot v. \end{aligned}$$

Similarly, if $f_{t-1,1}(v) > \psi^{(2)}$ and therefore $\delta_{\mathrm{u}}(\gamma, R) > \frac{1}{1-d}$, we have

$$J'_{t-1}(v) = \alpha_{+,12} \cdot v.$$

Now, assume $\frac{p^*}{p}\alpha_+^{(t)} > \frac{1-p^*}{1-p}\alpha_-^{(t)}$, i. e. the function is decreasing on $(-\infty, \psi^{(1)})$. If the lower bound of $D(v)$ is not smaller than $\psi^{(1)}$, which is equivalent to

$$f_{t-1,2}(v) \geq \psi^{(1)} \;\Leftrightarrow\; \delta_{\mathrm{l}}(\gamma, R) \leq \frac{1}{u-1},$$

the function is increasing on $D(v)$ and we have again that $f_{t-1,1}(v)$ is an optimal value for a. On the other hand, if $f_{t-1,2}(v) < \psi^{(1)}$, we have to compare the value of the function at $f_{t-1,2}(v)$ and at $f_{t-1,1}(v)$ to find the larger one. Again, there are two possibilities for $f_{t-1,1}(v)$ depending on the fact whether the inequality in (2.20) holds or not. The value of the function at $f_{t-1,2}(v)$ becomes

$$\begin{aligned}
&p\alpha_-^{(t)}(v + f_{t-1,2}(v)(u-1)) + (1-p)\alpha_+^{(t)}(v + f_{t-1,2}(v)(d-1)) \\
&= \big(p\alpha_-^{(t)} + (1-p)\alpha_+^{(t)} + \delta_1(\gamma, R)\big(\alpha_+^{(t)}(1-p)(1-d) - \alpha_-^{(t)}p(u-1)\big)\big) \cdot v \\
&= \alpha_{+,2}^{(t-1)} \cdot v,
\end{aligned}$$

so that we have to compare $\alpha_{+,2}^{(t-1)}$ with $\alpha_{+,1}^{(t-1)}$ and obtain for the value function $J'_{t-1}(v) = \max\{\alpha_{+,1}^{(t-1)}, \alpha_{+,2}^{(t-1)}\} \cdot v$. Combining the cases, it holds for $v \geq 0$

$$J'_{t-1}(v) = \alpha_+^{(t-1)} \cdot v = \alpha_+^{(t-1)} \cdot v^+. \tag{2.21}$$

We have also shown that $f_{t-1}^*(v)$ defined by the algorithm is an optimal value for a.

If $v < 0$, the proof works similarly and we can indeed show that

$$J'_{t-1}(v) = \alpha_-^{(t-1)} \cdot v = -\alpha_-^{(t-1)} \cdot v^-. \tag{2.22}$$

One only has to observe that now $\psi^{(2)} < 0 < \psi^{(1)}$ and that the slope on the interval $[\psi^{(2)}, \psi^{(1)}]$ is

$$\alpha_-^{(t)} \underbrace{s(u-d)\big(p(1-p^*) - (1-p)p^*\big)}_{>0},$$

hence the function to be maximized can also be decreasing on this interval, depending on the sign of $\alpha_-^{(t)}$. This fact makes it necessary to consider one more case as it is done in the algorithm. Putting (2.21) and (2.22) together completes the induction step and therefore also the proof. □

Interpretation of the optimal policy The structure of the optimal policy is similar to the model without shortselling. The only difference is that we now might have to invest more than the current wealth into to the risky asset or even have to invest a negative amount. Furthermore, this factor depends again only on the sign of the amount of wealth and not on its level. However, this selection is not constant over time. The "sequence of fractions" is deterministic and obtained by the algorithm above. It depends on the interplay of the chosen risk level R and the safety level γ with p and p^*.

We conclude this section and chapter by giving an example on the structure of the occurring parameters.

Example 2.1. We assume that $p^* < \frac{1}{2}$ and $p \geq 2p^*$. The first inequality is equivalent to

$$\frac{1}{1-d} > \frac{1}{u-1}. \tag{2.23}$$

Furthermore, let $R = 1$ and $\gamma \geq \max\{p, 1-p\}$. It follows from (2.4) and the definition of $\delta_{\mathrm{u}}(\gamma, R)$ and $\delta_{\mathrm{l}}(\gamma, R)$ that

$$\delta_{\mathrm{u}}(\gamma, R) = \frac{1}{1-d}, \quad \delta_{\mathrm{l}}(\gamma, R) = \frac{1}{u-1}.$$

We claim that

$$\alpha_+^{(t)} = \left(\frac{p}{p^*}\right)^{T-t} > 0, \quad \alpha_-^{(t)} = \left(2 - \frac{p}{p^*}\right)\left(1 + \frac{p-p^*}{1-p^*}\right)^{T-t-1} \leq 0, \quad t = 0, 1, \ldots, T-1.$$

For $t = T-1$ it holds by the algorithm

$$\alpha_+^{(T-1)} := 1 + (u-d)(p-p^*)\frac{1}{1-d} = \frac{p}{p^*}$$

and

$$\alpha_-^{(T-1)} := 1 - (u-d)(p-p^*)\frac{1}{1-d} = 2 - \frac{p}{p^*}.$$

Now assume that the assertion holds for $t \in \{1, \ldots, T-1\}$. First,

$$\begin{aligned}\alpha_+^{(t-1)} &= \alpha_{+,11} = \alpha_+^{(t)} + \delta_{\mathrm{u}}(\gamma, R)(u-d) \cdot \alpha_+^{(t)} \cdot \left(p(1-p^*) - (1-p)p^*\right)\\ &= \alpha_+^{(t)}\left(1 + \frac{u-d}{1-d} \cdot (p-p^*)\right) = \left(\frac{p}{p^*}\right)^{T-t}\left(\frac{p}{p^*}\right) = \left(\frac{p}{p^*}\right)^{T-t+1}.\end{aligned}$$

Because of $\alpha_-^{(t)} \leq 0$ and (2.23) we have $\alpha_-^{(t-1)} = \max\{\alpha_{-,12}, \alpha_{-,21}\}$ by case (a) (ii) from the algorithm. But indeed it is easily verified by a straightforward calculation that $\alpha_{-,21} > \alpha_{-,12}$ under the given assumption $p^* < \frac{1}{2}$. Hence, we obtain

$$\begin{aligned}\alpha_-^{(t-1)} &= \alpha_-^{(t)} - \delta_{\mathrm{l}}(\gamma, R)(u-d) \cdot \alpha_-^{(t)} \cdot \left((1-p)p^* - p(1-p^*)\right)\\ &= \alpha_-^{(t)} - \frac{u-d}{u-1} \cdot \alpha_-^{(t)} \cdot (p^* - p)\\ &= \left(2 - \frac{p}{p^*}\right)\left(1 + \frac{p-p^*}{1-p^*}\right)^{T-t-1}\left(1 + \frac{p-p^*}{1-p^*}\right)\\ &= \left(2 - \frac{p}{p^*}\right)\left(1 + \frac{p-p^*}{1-p^*}\right)^{T-t}.\end{aligned}$$

Consequently, with the parameters specified as above Theorem 2.7 yields that the optimal value of (UM') is given by

$$J_0(v) = \left(\frac{p}{p^*}\right)^T \cdot v^+ - \left(2 - \frac{p}{p^*}\right)\left(1 + \frac{p-p^*}{1-p^*}\right)^{T-1} \cdot v^-, \quad v \in S.$$

For $t \in \{0, 1, \ldots, T-1\}$, the (unique) optimal one step policy f_t^* is given by $f_t^*(v) = f^{(1)}(v)$ if $v > 0$ and $f_t^*(v) = f^{(2)}(v)$ if $v \leq 0$.

3 Dynamic risk measures

In this chapter the notion of *dynamic risk measures* is introduced as a natural generalization of static risk measures. Let us first describe this construction and review the related literature.

3.1 An overview on the literature

Static risk measurement deals with the quantification of a random financial position at a future date T by a real number. This value is interpreted as the risk of the position measured at time 0. In Chapter 1, we have described how the seminal paper Artzner et al. (1999) generated a vast growth of the literature dealing with theoretical and practical aspects of such (static) risk measures. Naturally, the development of representations and applications of dynamic risk measures followed with some delay. However, there is one working paper by Wang which is from the same year as the publication of the aforementioned work (1999). To our knowledge, Wang was the first one to use the term *dynamic risk measure* which already appeared in an early draft of his paper in 1996.

His work on a class of dynamic risk measures includes a short motivation for the introduction of dynamic risk measures based on three main reasons. The first one comes from the necessity to generalize the static setting to a framework where processes can be considered. This is due to the fact that companies often face risk that consists of different uncertain values of financial positions that evolve over time, imagine for example intermediate cash flows. Consequently, it is desirable to have risk measures for processes. The second generalization of the static model follows from the fact that over certain periods of time, companies might be willing to adapt the measured risk when more information becomes available. More precisely, at each point of time t before the end of the considered time framework $T \in (0, \infty]$, the value of the risk measure should be an $\mathcal{F}_t$-measurable random variable if the information is modelled by some filtration $(\mathcal{F}_t)_{t \in \mathcal{T}}$, where $\mathcal{T} = \{0, 1, \ldots, T\}$ or $\mathcal{T} = [0, T]$. Furthermore, this adaption process should be consistent in the sense that there is a certain connection between the risk measured at different points of time. We will later formalize this. As a last reason, optimization problems are often of a dynamic structure which then yields a demand for dynamic measures which have to be incorporated in the optimization problem. In this chapter, we mainly concentrate on dealing with the first two aspects. In the following two chapters though, we will see how dynamic risk measures themselves can be generated by an economically motivated optimization problem.

Let us now have a more thorough look on the literature. Most of the relevant works distinguish dynamic risk measure by these aspects:

- Dynamic risk measures can be introduced only for *final payments* or for *processes*.
- In the latter case, *bounded* or *unbounded* processes are distinguished.
- Analogously to the static case, *convex* or *coherent* dynamic risk measures can be investigated.
- The time–framework can be *discrete* or *continuous* and *finite* or *infinite*.
- Some works only deal with *finite probability spaces*, others with *general* ones.

In this work, we focus on coherent dynamic risk measures for unbounded processes in a finite and discrete time framework with no restrictions on the probability space. Nonetheless, we now give an overview on the existing literature in all cases.

The aforementioned work by Wang (1999) describes an iterated version of the static Value–at–Risk. In a finite and discrete time framework, the risk at a fixed point of time is calculated by measuring the remaining risk after the next period based on the available information and adding this term to the value of the financial position in the following period. Taking the Value–at–Risk of this sum conditioned on the current information yields the value of the risk measure at this fixed date. As a generalization of this example, the class of likelihood–based risk measures is introduced and characterized by four properties which are all based on the dynamic structure of the risk measures. However, no actual dynamic version of the classic coherence properties is used, which is why we will not further deal with this work. It is interesting in itself, but not linked to the settings introduced in all the following works. Furthermore, the Value–at–Risk is not a coherent risk measure, so a dynamic version of it is nothing we want to investigate further. However, in Hardy and Wirch (2004) a similar approach is used to define a so-called iterated Conditional Tail Expectation (CTE), where the risk is measured only for final payments. This is done simply by taking at every point of time the conditional expectation of the future risk.

Another early work is the one by Cvitanić and Karatzas (1999). Based on an early draft of Artzner et al. (1999), the static approach is generalized to a setting in continuous and finite time. Here, a classic financial market with a final financial position that has to be hedged is considered. Actually, the work is not dealing with dynamic risk measures since the risk is introduced only at time $t = 0$ (so–called *initial risk measurement*). It is defined as a classic max–min approach over a set of probability measures and a set of admissible strategies respectively where the target value to be optimized is the expectation of the discounted shortfall that occurs if a given liability can not be hedged. Because of the continuous time–framework, this paper is not relevant for our work. However, due to its early publication, it was worth to be mentioned. A similar investigation in discrete time is given in Boda and Filar (2006), where a dynamic version of the Average Value-at-Risk at a point of time is just defined as the static Average Value-at-Risk of the future random loss of a portfolio when using a fixed portfolio strategy. Using this risk measure, optimization problems are considered and time-consistency then is defined using the well-known principle of optimality.

A more straightforward generalization approach of coherent static risk measures was followed by some of the same authors of Artzner et al. (1999), merely in Artzner et al. (2004) and Delbaen (2006). In the former work, the static representation result (compare Theorem 1.1) is generalized to measuring the initial risk of discrete- and infinite-time processes simply by an extension of the probability space. In finite time, risk measurement over time is then investigated by introducing so-called risk-adjusted value processes in terms of generalized Snell envelopes. In contrast to works described below, expectations are not taken of discounted sums of the components of processes but rather of the component of the process at a certain stopping time. This model is called the supremum case in Burgert (2005). The approach is more thoroughly dealt with in Delbaen (2006), adding some consistency requirement on the risk measurement via the notion of stable sets of probability measures (compare Section 3.3). A similar approach is introduced in Jobert and Rogers (2005) for a finite probability space using pricing operators. The authors reason that this simplifies the axiomatic approach.

In a sequence of papers, a comparable framework is investigated for a number of different models. The functionals that are characterized are called monetary risk measures, which corresponds to the fact that they are non-decreasing rather than non-increasing on their domain (contrary to the functionals that we just call risk measures). In Cheridito et al. (2006), such functionals are introduced and characterized for bounded processes in finite and infinite discrete-time. A main focus is again on the property called time-consistency. This means that the same risk of a financial position should result "irrespective of whether it is calculated directly or in two steps backwards in time". In Cheridito and Kupper (2006), this model is supplemented by considering dynamic monetary risk measures that are composed of static ones. Representation theorems for coherent and convex risk measures for bounded and unbounded processes in continuous time can be found in Cheridito et al. (2004) and Cheridito et al. (2005) respectively.

In discrete and finite time, which is the main subject of this thesis, there are some more interesting works. By the characterization theorems given there, calculating the risk of a process at every point of time is done by applying a static risk measure to the discounted sum of the future values of the process. This is why these works deal in fact with dynamic risk measures for final payments. However, they yield quite a practical view on dynamic risk measure, hence we consider them important. The dynamic risk measure that we will describe in the following chapter resembles a typical element of these classes but has the advantage not to depend only on the discounted sum of the components of a process of financial positions. Furthermore, we will see that our dynamic risk measure is not law invariant, a property that is not always desirable as shown in the simple Example 4.3. Contrary to this, in Weber (2006), law invariant dynamic risk measures for bounded processes are investigated. They are only characterized by the properties of monotonicity and translation invariance and shown to be at every point of time a static risk measure of the conditional distribution of the discounted sum of the future values of a process. A similar result, but without the assumption of law invariance, is given in Riedel (2004) for coherent dynamic risk measures. Here, again the risk of the discounted sum of the future values of a process is measured at every point of time, but now with a conditional static risk measure. Compare Section 1.4 for some notes on the distinction of these two notions. One drawback in Riedel's work is the assumption of a finite probability space. This is overcome in Burgert (2005), where the result is generalized to bounded and unbounded processes on general probability spaces. It is merely obtained as a consequence of more general characterization theorems of convex and coherent dynamic risk measures for final payments in continuous and discrete time. It is shown, similar to Delbaen (2006), that there is a correspondence between a dynamic risk measure and a stable set of probability measures, if a certain time–consistency property is imposed. Complementary, a similar characterization, though in a different context, is given in Epstein and Schneider (2003) for utility functions by using recursive multiple priors. We will describe these results more formally in the following section and also introduce an interesting translation invariance property that is more general than the one used in the aforementioned works and which is first used in Frittelli and Scandolo (2006). In this work, a class of risk measures for processes in discrete and finite time is investigated. The risk measurement takes only place at time 0.

For the sake of completeness, a few other works for processes in discrete and finite time are worth to be mentioned. Based on the earlier work Roorda et al. (2005), dynamic risk measures for final payments on a finite probability space are characterized in Roorda and Schumacher (2007) for different notions of time–consistency. Furthermore, these are all compared, and for two of them, a consistent version of the Average Value–at–Risk is given, hereby completing a note in Artzner et al. (2004) which shows that the definition of a dynamic Average Value–at–Risk in the latter work is not consistent. In Section 4.4, we will give another example to overcome this drawback by using a stable set of probability measures.

In Jouini et al. (2004), a totally different setting for bounded and d–dimensional portfolios is used. The risk measure now has values in $\mathbb{R}^n$ with $n \leq d$. This reduction is based on aggregation procedures. One–dimensional–valued risk measures for portfolio vectors are also considered in Burgert and Rüschendorf (2006), where a main focus is put on the relationship with stochastic orderings and dependence concepts. In Bion-Nadal (2006b), dynamic risk measures in discrete and continuous infinite time are characterized using conditional static risk measures on a larger space. In continuous time, an approach via BMO martingales[1] is taken in Bion-Nadal (2006a). Finally, convex dynamic risk measures for bounded processes in discrete and finite time are again investigated in Föllmer and Penner (2006) by using the standard notion of conditional static risk measures.

We conclude that there is a number of publications which deal with dynamic risk measures for final payments or discounted sums of future payments (in a discrete time framework). An important property of such measures is time-consistency, which can be defined in various ways. But this is not a main subject in this work. Since we want to introduce a dynamic risk measure in discrete and finite time and investigate its properties in the following chapter, we now briefly summarize the most relevant results that can be related to our dynamic risk measure. We focus on the accordant notion of time-consistency which is related to stability of sets of probability measures and thoroughly deal with the property of translation invariance.

3.2 Definitions, axioms and properties

In this section, our aim is to give a short overview on properties of dynamic risk measures in discrete and finite time which we consider important. Hence, let the time set be $\mathcal{T} = \{0, 1, \dots, T\}$ with $T \in \mathbb{N}$ and let $(\mathcal{F}_t)_{t\in\mathcal{T}}$ be a filtration on a given probability space $(\Omega, \mathcal{F}, \mathbb{P})$. A dynamic risk measure for an $(\mathcal{F}_t)_{t\in\{0,1,\dots,T\}}$-adapted process then is a mapping that assigns at every point of time $t \in \{0, 1, \dots, T-1\}$ an $\mathcal{F}_t$-measurable risk to the process, such that the sequence of the values of the risk measures is again an $(\mathcal{F}_t)_{t\in\{0,1,\dots,T\}}$-adapted process. Furthermore, the dynamic risk measure should satisfy two elementary properties, which are given in the following definition. Let

$$\mathcal{X} := \{(I_1, \dots, I_T) \mid I_t \in \mathcal{X}_t,\ t = 1, \dots, T\},$$

be the space of all integrable and $(\mathcal{F}_t)_{t\in\mathcal{T}}$-adapted processes, where

$$\mathcal{X}_t := L^1(\Omega, \mathcal{F}_t, \mathbb{P}), \quad t = 0, 1, \dots, T.$$

For convenience, we enlarge the random vector $(I_1, \dots, I_T)$ by a dummy component $I_0 \equiv 0$ whenever necessary. All occurring equalities and inequalities between random variables are understood in the $\mathbb{P}$-almost sure sense.

Definition 3.1. *Let $\rho : \Omega \times \{0, 1, \dots, T-1\} \times \mathcal{X} \to \overline{\mathbb{R}}$ be a mapping and set $\rho_t(I)(\omega) := \rho(\omega, t, I)$ for all $(\omega, t, I) \in \Omega \times \{0, 1, \dots, T-1\} \times \mathcal{X}$. Then ρ is called a* dynamic risk measure *if $(\rho_t(I))_{t=0,1,\dots,T-1}$ is an $(\mathcal{F}_t)_{t=0,1,\dots,T-1}$-adapted process and if the following two conditions are fulfilled:*

[1] A BMO martingale is a uniformly integrable martingale starting in 0 that fulfills a certain additional integrability condition which uses stopping times.

(IOP) ρ is *independent of the past*, i. e. for every $t = 0, 1, \dots, T-1$ and $I \in \mathcal{X}$, $\rho_t(I)$ does not depend on $I_1, \dots, I_{t-1}$.

(MON) ρ is *monotone*, i. e. for all $I^{(1)}, I^{(2)} \in \mathcal{X}$ with $I_t^{(1)} \leq I_t^{(2)}$ for all $t = 1, \dots, T$ it holds

$$\rho_t(I^{(1)}) \geq \rho_t(I^{(2)}), \quad t = 0, 1, \dots, T-1.$$

Remark. 1. We sometimes identify ρ with the corresponding sequence of mappings $(\rho_t)_{t=0,1,\dots,T-1}$.

2. Condition (IOP) means that for every $I \in \mathcal{X}$ it holds

$$\rho_t(I_1, \dots, I_T) = \rho_t(0, \dots, 0, I_t, \dots, I_T), \quad t = 0, 1, \dots, T-1.$$

In the following, we further assume that there exists a constant interest rate $r > -1$. By $e_k \in \mathcal{X}$ we denote the income process that is one at a fixed point of time $k \in \{1, \dots, T\}$ and zero otherwise. Let us now review the most important properties of dynamic risk measures. Afterwards, we introduce the notion of coherence, analogously to the static setting.

Translation invariance properties The economic interpretation of the values of the risk measured is based on these properties. Let ρ be a dynamic risk measure.

(TI1) For all $I \in \mathcal{X}$, $t \in \{0, 1, \dots, T-1\}$ and $Z \in L^1(\Omega, \mathcal{F}_t, \mathbb{P})$ it holds

$$\rho_t(I + Z \cdot e_k) = \rho_t(I) - \frac{Z}{(1+r)^{k-t}}, \quad k = t, \dots, T.$$

(TI2) This is (TI1) in the case $k = T$, i. e. for all $Z \in L^1(\Omega, \mathcal{F}_t, \mathbb{P})$ and $I \in \mathcal{X}$ it holds

$$\rho_t(I + Z \cdot e_T) = \rho_t(I) - \frac{Z}{(1+r)^{T-t}}.$$

(TI3) Let $t \in \{0, 1, \dots, T-1\}$. For every process

$$Z = (0, \dots, 0, Z_t, \dots, Z_T) \in \mathcal{X}$$

such that $\sum_{k=t}^{T} \frac{Z_k}{(1+r)^{k-t}}$ is $\mathcal{F}_t$-measurable it holds for all $I \in \mathcal{X}$

$$\rho_t(I + Z) = \rho_t(I) - \sum_{k=t}^{T} \frac{Z_k}{(1+r)^{k-t}}. \tag{3.1}$$

We obtain from (TI1) or (TI3)

$$\rho_t(I + \rho_t(I) \cdot e_t) = \rho_t(I) - \rho_t(I) = 0,$$

which provides the same economic interpretation of the value $\rho_t(I)$ as in the static setting.

Property (TI1) is an essential tool in Riedel (2004), while (TI2) is used in Weber (2006). (TI3) is inspired by Definition 4.7 in Frittelli and Scandolo (2006). We believe that it is the most natural extension of the static translation invariance in the context of Theorems 3.1 and 3.2, because in the one period setting, an $\mathcal{F}_0$-measurable, therefore predictable amount c is added to the $\mathcal{F}_T$-measurable random payoff at time T. Hence, in a multiperiod setting, the risk manager should be allowed to add at every time t a process Z to the income process I and in this manner diminish the risk, as long as the discounted sum of the future values of Z is known at time t. Otherwise, Equation (3.1) would not make sense.

However, considering a generalization of a risk measure by Pflug and Ruszczyński in the following chapter, we will see that a restriction to Z being a predictable process can be reasonable too, compare Proposition 4.3.

Coherence properties Analogously to the static case, dynamic risk measures can be distinguished by the properties of convexity and subadditivity. Furthermore, homogeneity plays a crucial role when connecting the two former properties.

(CVX) A dynamic risk measure ρ is called *convex* if for all $I^{(1)}, I^{(2)} \in \mathcal{X}$, $t \in \{0, 1, \ldots, T-1\}$ and $\Lambda \in L^\infty(\Omega, \mathcal{F}_t, \mathbb{P})$ with $0 \leq \Lambda \leq 1$ it holds

$$\rho_t(\Lambda I^{(1)} + (1-\Lambda) I^{(2)}) \leq \Lambda \rho_t(I^{(1)}) + (1-\Lambda)\rho_t(I^{(2)}).$$

(SUB) A dynamic risk measure ρ is called *subadditive* if for all $I^{(1)}, I^{(2)} \in \mathcal{X}$ and $t \in \{0, 1, \ldots, T-1\}$ it holds

$$\rho_t(I^{(1)} + I^{(2)}) \leq \rho_t(I^{(1)}) + \rho_t(I^{(2)}).$$

(HOM) A dynamic risk measure ρ is called *homogeneous* if for all $I \in \mathcal{X}$, $t \in \{0, 1, \ldots, T-1\}$ and $\Lambda \in L^\infty(\Omega, \mathcal{F}_t, \mathbb{P})$ with $\Lambda \geq 0$ it holds

$$\rho_t(\Lambda I) = \Lambda \rho_t(I).$$

As in the one-period model, under homogeneity, subadditivity and convexity are equivalent. The interpretation of the properties is the same as in the static model.

Consistency properties These properties are crucial tools when deriving representation theorems for dynamic risk measures.

(TCS) A dynamic risk measure ρ is called *time-consistent* if for all stopping times σ, τ on $\{0, 1, \ldots, T\}$ with $\sigma \leq \tau$, all processes $I \in \mathcal{X}$ and $Z \in L^1(\Omega, \mathcal{F}_\tau, \mathbb{P})$ it holds

$$\rho_\sigma(I + Z \cdot e_\tau) = \rho_\sigma(I + (1+r)^{T-\tau} Z \cdot e_T).$$

(DCS) A dynamic risk measure ρ is called *dynamically consistent* if for all processes $I^{(1)}, I^{(2)} \in \mathcal{X}$ with $I_t^{(1)} = I_t^{(2)}$ for a $t \in \{0, 1, \ldots, T-1\}$ it holds

$$\rho_{t+1}(I^{(1)}) = \rho_{t+1}(I^{(2)}) \quad \Rightarrow \quad \rho_t(I^{(1)}) = \rho_t(I^{(2)}).$$

Property (TCS) is used in Weber (2006) and Burgert (2005), whereas (DCS) is defined and motivated in Riedel (2004).

Technical properties To derive characterization theorems, usually some technical assumptions have to be made.

(REL) A dynamic risk measure ρ is called *relevant* if for all $A \in \mathcal{F}_T$ with $\mathbb{P}(A) > 0$ it holds

$$\mathbb{P}(\rho_t(-\mathbb{1}_A \cdot e_T) > 0) > 0, \quad t = 0, 1, \ldots, T-1.$$

(FAT) A dynamic risk measure ρ has the *Fatou-property*, if for every point of time $t \in \{0, 1, \ldots, T-1\}$ and all processes $I^{(n)} \in \mathcal{X}$, $n \in \mathbb{N} \cup \{\infty\}$, such that

$$\sup_{l \geq t} \mathbb{E}\,|I_l^{(n)}| \leq 1, \quad n \in \mathbb{N} \cup \{\infty\},$$

and $\sup_{l \geq t} |I_l^{(n)} - I_l^{(\infty)}| \xrightarrow{L^1} 0$, $n \to \infty$, it holds

$$\rho_k(I^{(\infty)}) \leq \liminf_{n \to \infty} \rho_k(I^{(n)}), \quad k \in \{t, \ldots, T-1\}.$$

(LIN) Let ρ be a dynamic risk measure. Then $a = (a_t)_{t=0,1,\dots,T-1}$ defined via

$$a_t(I) = \mathbb{1}_{(-\infty,0]}(\rho_t(I)), \quad I \in \mathcal{X}^\infty,$$

is called the *acceptance indicator*, where $\mathcal{X}^\infty \subset \mathcal{X}$ is the set of all bounded and adapted processes. Denote with $\mathcal{M}_c$ the set of all probability measures on the real line with compact support. Then ρ is called *law invariant* if there exists a measurable mapping $H_t : \mathcal{M}_c \to \{0,1\}$ such that for $t = 0,1,\dots,T-1$, $I \in \mathcal{X}^\infty$

$$a_t(I) = H_t(\mathcal{L}(I_T \mid \mathcal{F}_t)) \quad \mathbb{P} \text{ almost surely.}$$

Since there is a variety of proposals for the property of translation invariance, we briefly investigate the relationship of (TI1), (TI2) and (TI3). Obviously, (TI2) follows from (TI3). But indeed, introducing a weaker form of time-consistency, the following statement can be made.

Proposition 3.1. *Let ρ be a dynamic risk measure. Then the property (TI3) is fulfilled if (TI2) and*

(WCS) Let $t \in \{0,1,\dots,T-1\}$ and $k \in \{t,\dots,T\}$. Then it holds for all $Z \in L^1(\Omega, \mathcal{F}_k, \mathbb{P})$ and $I \in \mathcal{I}$

$$\rho_t(I + Z \cdot e_k) = \rho_t(I + (1+r)^{T-k} Z \cdot e_T)$$

hold. Furthermore, (TI3) implies (TI1) and a weaker form of (WCS), namely if only $Z \in L^1(\Omega, \mathcal{F}_t, \mathbb{P})$ is allowed.

Proof. Assume that (TI2) and (WCS) hold. Take $Z = (0,\dots,0,Z_t,\dots,Z_T) \in \mathcal{X}$ such that $\sum_{k=t}^T \frac{Z_k}{(1+r)^{k-t}}$ is $\mathcal{F}_t$ measurable. Then for $I \in \mathcal{X}$

$$\begin{aligned}\rho_t(I+Z) &= \rho_t(I + Z_t \cdot e_t + \dots + Z_T \cdot e_T) \overset{\text{(WCS)}}{=} \rho_t\left(I + \sum_{k=t}^T (1+r)^{T-k} Z_k \cdot e_T\right) \\ &= \rho_t\Bigg(I + (1+r)^{T-t} \underbrace{\sum_{k=t}^T \frac{Z_k}{(1+r)^{k-t}}}_{\mathcal{F}_t\text{-measurable}} \cdot e_T\Bigg) \overset{\text{(TI2)}}{=} \rho_t(I) - \sum_{k=t}^T \frac{Z_k}{(1+r)^{k-t}}.\end{aligned}$$

The second part is almost trivial, since (TI3) obviously implies (TI1). To show (WCS) for $Z \in L^1(\Omega, \mathcal{F}_t, \mathbb{P})$ let $I \in \mathcal{I}$. Then

$$\begin{aligned}\rho_t(I + Z \cdot e_k) &\overset{\text{(TI3)}}{=} \rho_t(I) - \frac{Z}{(1+r)^{k-t}} = \rho_t(I) - \frac{(1+r)^{T-k} Z}{(1+r)^{T-t}} \\ &\overset{\text{(TI3)}}{=} \rho_t(I + (1+r)^{T-k} Z \cdot e_T).\end{aligned}$$

□

Remark. We call the property (WCS) *weak consistency*. Obviously, (TCS) implies (WCS).

The following corollary shows the equivalence of the three translation invariance properties under weak consistency.

Corollary 3.1. *Let (WCS) hold. Then (TI1), (TI2) and (TI3) are all equivalent.*

Proof. Obviously, (TI3) implies (TI1), from which again (TI2) is a special case. Consequently, the following sequence of implications is valid:

$$\text{(TI3)} \quad \Rightarrow \quad \text{(TI1)} \quad \Rightarrow \quad \text{(TI2)} \quad \overset{\text{Proposition 3.1}}{\Rightarrow} \quad \text{(TI3)}.$$

□

We will later see how the previous corollary implies that the following definition of (dynamic) coherence is in accordance with the definitions used in related works, e. g. Weber (2006) or Burgert (2005).

Definition 3.2. *A dynamic risk measure ρ is called* coherent, *if it fulfills (TI3), (SUB) and (HOM), i. e. if it is translation invariant (of the third kind), subadditive and homogeneous.*

To state representation theorems for dynamic risk measure, we need the following definition, used e. g. in Artzner et al. (2004), Definition 3.1. If $\mathcal{Q}$ is a set of probability measures with $\mathbb{P} \in \mathcal{Q}$, we denote with $\mathcal{Q}^e$ the subset of all probability measures which are equivalent to $\mathbb{P}$ and with $\mathcal{Q}^*$ all the ones which are absolutely continuous with respect to $\mathbb{P}$. We usually only consider such sets where $\mathcal{Q} = \mathcal{Q}^*$.

If Q is a probability measure with $Q \ll \mathbb{P}$ on $\mathcal{F}_T$ and $L_T^Q := \frac{\mathrm{d}Q}{\mathrm{d}\mathbb{P}}$ the resulting density, we are able to introduce the so-called *density process* of Q with respect to $\mathbb{P}$ via

$$L_t^Q := \mathbb{E}\,[L_T^Q \,|\, \mathcal{F}_t], \quad t = 0, 1, \ldots, T.$$

It is well-known (see Attachment A), that L_t^Q is the density of Q with respect to $\mathbb{P}$ on $\mathcal{F}_t$ for every $t \in \{0, 1, \ldots, T\}$. Obviously, the process $(L_t^Q)_{t=0,1,\ldots,T}$ is an $(\mathcal{F}_t)_{t=0,1,\ldots,T}$-martingale.

Definition 3.3. *A set $\mathcal{Q}$ of probability measures is called* stable (under pasting), *if for all $Q_1, Q_2 \in \mathcal{Q}^e$ with density processes $(L_t^{Q_1})_{t=0,1,\ldots,T}$, $(L_t^{Q_2})_{t=0,1,\ldots,T}$ and all stopping times $\tau \in \{0, 1, \ldots, T\}$ the process*

$$L_t^{(\tau)} := \begin{cases} L_t^{Q_1} & , \quad t = 0, 1, \ldots \tau, \\ L_\tau^{Q_1} \frac{L_t^{Q_2}}{L_\tau^{Q_2}} & , \quad t = \tau + 1, \ldots, T, \end{cases}$$

defines an element $Q^{(\tau)} \in \mathcal{Q}$ which is called the pasting of Q_1 and Q_2 in τ.

Theorem 3.1. *A mapping $\rho : \Omega \times \{0, 1, \dots, T-1\} \times \mathcal{X} \to \overline{\mathbb{R}}$ is a relevant and time-consistent coherent dynamic risk measure with the Fatou property if and only if there exists a stable, convex and $L^\infty(\Omega, \mathcal{F}_T, \mathbb{P})$-closed set $\mathcal{Q}$ of probability measures that are absolutely continuous with respect to $\mathbb{P}$ with $\frac{\mathrm{d}Q}{\mathrm{d}\mathbb{P}} \in L^\infty(\Omega, \mathcal{F}_T, \mathbb{P})$, $Q \in \mathcal{Q}$, such that for all $t = 0, 1, \dots, T-1$*

$$\rho_t(I) = \operatorname{ess.\,sup}_{Q \in \mathcal{Q}} \mathbb{E}_Q \left[-\sum_{k=t}^{T} \frac{I_k}{(1+r)^{k-t}} \,\Big|\, \mathcal{F}_t \right], \quad I \in \mathcal{X}.$$

Proof. This is Theorem 3.27 in Burgert (2005) in the case $p = 1$ and $q = \infty$. Note that in the cited work, (TI1) instead of (TI3) is assumed in the definition of coherence. Since time consistency implies weak consistency and because of Corollary 3.1, the theorem is still valid when using our notion of coherence. □

Remark. The same representation is derived in Riedel (2004) if $|\Omega| < \infty$. One only has to replace time-consistency by dynamic consistency and one does not need the Fatou property. Furthermore, the notion of stability of sets of probability measures is called *consistency*. We will later show that the definition in Riedel (2004) is indeed nothing else than stability.

In Weber (2006), only bounded processes are considered. Hence, let

$$\mathcal{X}^\infty := \{(I_1, \dots, I_T) \mid I_t \in L^\infty(\Omega, \mathcal{F}_t, \mathbb{P}),\ t = 1, \dots, T\}.$$

Theorem 3.2. *A dynamic risk measure $\rho : \Omega \times \{0, 1, \dots, T-1\} \times \mathcal{X}^\infty \to \overline{\mathbb{R}}$ satisfies (TI2), (TCS) and (DIN) if and only if there exists a (unique) sequence $(\Theta_t)_{t=0,1,\dots,T-1}$ of monotone, translation and law invariant static risk measures such that*

$$\rho_t(I) = \Theta_t\Big(\mathcal{L}\Big(-\sum_{k=t}^{T} \frac{I_k}{(1+r)^{k-t}} \,\Big|\, \mathcal{F}_t\Big)\Big), \quad I \in \mathcal{X}^\infty.$$

Proof. This is Theorem 4.6 in Weber (2006). There it is formulated only for final payments, but by time-consistency, the slightly more general case mentioned here also holds. □

Remark. In a further step in the mentioned work, another consistency property is introduced in order to get $\Theta_t = \Theta_0$, $t = 0, 1, \dots, T-1$.

The aim of this section was to introduce properties of dynamic risk measures and in particular to investigate the different notions of translation invariance. Furthermore, we have seen how these properties can be used to derive representation theorems for dynamic risk measures under different collections of properties. In the following chapter, we introduce a concrete dynamic risk measure and state its relevant properties. We will also see how this risk measure fits with the two aforementioned theorems. Before doing so, let us conclude this chapter by first generalizing the notion of stability of sets of probability measures and then show its equivalence to what is called consistency of such sets. This is also related to a supermartingale property of dynamic risk measures for final payments.

3.3 Stable sets of probability measures

Let the reference probability measure $\mathbb{P}$ and $k \in \{1, \dots, T\}$ be fixed throughout the first part of this subsection.

If Q is another probability measure with $Q \ll \mathbb{P}$ on $\mathcal{F}_k$ and $L_k^Q := \frac{\mathrm{d}Q}{\mathrm{d}\mathbb{P}}\big|_{\mathcal{F}_k}$ the resulting density, we are able to introduce the so-called *density process* of Q with respect to $\mathbb{P}$ via

$$L_t^Q := \mathbb{E}\,[L_k^Q \,|\, \mathcal{F}_t], \quad t = 0, 1, \dots, k.$$

As above, L_t^Q is the density of Q with respect to $\mathbb{P}$ on $\mathcal{F}_t$ for every $t \in \{0, 1, \dots, k\}$ and the process $(L_t^Q)_{t=0,\dots,k}$ is an $(\mathcal{F}_t)_{t=0,\dots,k}$-martingale. We can now formulate a slight generalization of stability of sets of probability measures.

Definition 3.4. *A set $\mathcal{Q}$ of probability measures is called* stable (under pasting) on $\mathcal{F}_k$, *if for all $Q_1, Q_2 \in \mathcal{Q}^e$ with density processes $(L_t^{Q_1})_{t=0,\dots,k}, (L_t^{Q_2})_{t=0,\dots,k}$ and all stopping times $\tau \leq k$ the process*

$$L_t^{(\tau)} := \begin{cases} L_t^{Q_1} & , \quad t = 0, 1, \dots \tau, \\ L_\tau^{Q_1} \frac{L_t^{Q_2}}{L_\tau^{Q_2}} & , \quad t = \tau + 1, \dots, k, \end{cases}$$

defines an element $Q^{(\tau)} \in \mathcal{Q}$ which is called the pasting of Q_1 and Q_2 in τ on $\mathcal{F}_k$.

Remark. 1. For $k = T$ this is just the notion used e. g. in Artzner et al. (2004), compare our Definition 3.3. Consequently, we sometimes refer to stability on $\mathcal{F}_T$ just as stability.

2. Lemma 6.41 and Definition 6.42 in Föllmer and Schied (2004) provide a way how we can directly obtain the probability measure $Q^{(\tau)}$ from Q_1 and Q_2:

$$Q^{(\tau)}(A) = \mathbb{E}_{Q_1}[Q_2(A \,|\, \mathcal{F}_\tau)], \quad A \in \mathcal{F}_k.$$

(ii) For calculations it is sometimes helpful to notice that for all $Q \ll \mathbb{P}$ on $\mathcal{F}_k$ and stopping times ν, σ on $\{0, 1, \dots, k\}$ with $\nu \geq \sigma$ holds

$$\mathbb{E}_Q[Z \cdot L_\sigma^Q \,|\, \mathcal{F}_\sigma] = \mathbb{E}\left[Z \cdot L_\nu^Q \,|\, \mathcal{F}_\sigma\right], \quad Z \in L^1(\Omega, \mathcal{F}_\nu, \mathbb{P}). \tag{3.2}$$

The equality is understood in the almost sure sense with respect to $\mathbb{P}$ and therefore also with respect to Q.

In the next chapter we will need the following useful result. It is related to the corollary after Lemma 4.1 in Artzner et al. (2004) and a generalization of Lemma 3 in Detlefsen and Scandolo (2005), which is mentioned but not proved in that work. We give a detailed proof here.

Proposition 3.2. *Let $\mathcal{Q}$ be stable on $\mathcal{F}_k$ and $\nu \leq \sigma \leq k$ stopping times. Then for every $Q \ll \mathbb{P}$ (not necessarily in $\mathcal{Q}$)*

$$\mathbb{E}_Q\Big[\operatorname*{ess.sup}_{Q' \in \mathcal{Q}^e} \mathbb{E}_{Q'}[Z \,|\, \mathcal{F}_\sigma] \,\Big|\, \mathcal{F}_\nu\Big] = \operatorname*{ess.sup}_{Q' \in \mathcal{Q}^e} \mathbb{E}_Q[\mathbb{E}_{Q'}[Z \,|\, \mathcal{F}_\sigma] \,|\, \mathcal{F}_\nu], \quad Z \in L^1(\Omega, \mathcal{F}_k, \mathbb{P}).$$

Proof. Let $Z \in L^1(\Omega, \mathcal{F}_k, \mathbb{P})$. The essential tool is Lemma 5.2 in Delbaen (2006). By taking in their notation the process $X = (X_1, \dots, X_k) = (Z, \dots, Z)$ we obtain $X_\tau = Z$ for every stopping time $\tau \leq k$. We conclude that the set

$$\{\mathbb{E}_{Q'}[Z \,|\, \mathcal{F}_\sigma] \,|\, Q' \in \mathcal{Q}^e\}$$

is closed for taking maxima.

Consequently, by Theorem A.32 in Föllmer and Schied (2004) there exists a sequence $(Q_n)_{n\in\mathbb{N}} \subset \mathcal{Q}^e$ such that

$$\mathbb{E}_{Q_n}[Z \,|\, \mathcal{F}_\sigma] \uparrow \operatorname*{ess.sup}_{Q' \in \mathcal{Q}^e} \mathbb{E}_{Q'}[Z \,|\, \mathcal{F}_\sigma], \; n \to \infty.$$

This holds $\mathbb{P}$–almost surely and consequently also Q–almost surely. Since we have $\mathbb{E}_{Q_n}[Z \,|\, \mathcal{F}_\sigma] \geq \mathbb{E}_{Q_1}[Z \,|\, \mathcal{F}_\sigma]$ and $\mathbb{E}_Q[|\mathbb{E}_{Q_1}[Z \,|\, \mathcal{F}_\sigma]|] < \infty$ the conditional version of the monotone convergence theorem can be applied and we obtain

$$\begin{aligned}
\mathbb{E}_Q\Big[\operatorname*{ess.sup}_{Q' \in \mathcal{Q}^e} \mathbb{E}_{Q'}[Z \,|\, \mathcal{F}_\sigma] \,\Big|\, \mathcal{F}_\nu\Big] &= \mathbb{E}_Q[\lim_{n\to\infty} \mathbb{E}_{Q_n}[Z \,|\, \mathcal{F}_\sigma] \,|\, \mathcal{F}_\nu] \\
&= \lim_{n\to\infty} \mathbb{E}_Q[\mathbb{E}_{Q_n}[Z \,|\, \mathcal{F}_\sigma] \,|\, \mathcal{F}_\nu] \qquad (3.3) \\
&\leq \operatorname*{ess.sup}_{Q' \in \mathcal{Q}^e} \mathbb{E}_Q[\mathbb{E}_{Q'}[Z \,|\, \mathcal{F}_\sigma] \,|\, \mathcal{F}_\nu].
\end{aligned}$$

On the other hand, we have for all $Q_0 \in \mathcal{Q}^e$

$$\mathbb{E}_Q\Big[\operatorname*{ess.sup}_{Q' \in \mathcal{Q}^e} \mathbb{E}_{Q'}[Z \,|\, \mathcal{F}_\sigma] \,\Big|\, \mathcal{F}_\nu\Big] \geq \mathbb{E}_Q[\mathbb{E}_{Q_0}[Z \,|\, \mathcal{F}_\sigma] \,|\, \mathcal{F}_\nu],$$

so the reverse inequality in (3.3) also holds. This proves the assertion. □

Remark. The proof remains valid if the set $\{\mathbb{E}_{Q'}[Z \,|\, \mathcal{F}_\sigma] \,|\, Q' \in \mathcal{Q}^e\}$ is replaced with an arbitrary set of $\mathbb{P}$–integrable, $\mathcal{F}_\sigma$–measurable random variables which is closed for taking maxima.

To conclude this section, we prove that the notion of consistency, defined on finite probability spaces in Riedel (2004), is equivalent to stability. To this extend, let us first briefly describe the setting introduced in the aforementioned paper. From now on we assume $|\Omega| < \infty$ and that for the filtration it holds

$$\mathcal{F}_t = \sigma(Y_1, \dots, Y_t), \quad t = 1, \dots, T,$$

for some family of random variables $(Y_t)_{t=1,\dots,T}$ on $(\Omega, \mathcal{F}, \mathbb{P})$. Let

$$\mathcal{H}_t := \text{Range}((Y_1, \dots, Y_t)), \quad t = 1, \dots, T.$$

Furthermore, for any set $\mathcal{Q}$ of probability measures on $(\Omega, \mathcal{F})$, $t \in \{0, 1, \dots, T-1\}$ and $\xi_t \in \mathcal{H}_t$ let

$$\mathcal{Q}^{\xi_t} := \{Q(\cdot \mid (Y_1, \dots, Y_t) = \xi_t) \mid Q \in \mathcal{Q},\, Q((Y_1, \dots, Y_t) = \xi_t) > 0\}.$$

The following definition is introduced in Riedel (2004).

Definition 3.5. *Assume that for all $t = 0, 1, \dots, T-1$ and every history $\xi_t \in \mathcal{H}_t$, $\mathcal{Q}^{\xi_t}$ is a closed and convex set of (conditional) probability measures. The collection $(\mathcal{Q}^{\xi_t})_{\xi_t \in \mathcal{H}_t, t=0,1,\dots,T-1}$ is called a family of conditional probability measures.*

(i) Fix $t \in \{0, 1, \dots, T-1\}$ and $\xi_t \in \mathcal{H}_t$. Furthermore, choose for all histories $\xi_{t+1} = (\xi_t, y_{t+1}) \in \mathcal{H}_{t+1}$ a measure $Q^{(\xi_t, y_{t+1})} \in \mathcal{Q}^{(\xi_t, y_{t+1})}$ and $R^{\xi_t} \in \mathcal{Q}^{\xi_t}$. Then the composite probability measure *$Q^{(\xi_t, Y_{t+1})} R^{\xi_t}$ is defined via*

$$Q^{(\xi_t, Y_{t+1})} R^{\xi_t}(A) := \sum_{y_{t+1} : (\xi_t, y_{t+1}) \in \mathcal{H}_{t+1}} Q^{(\xi_t, y_{t+1})}(A) \cdot R^{\xi_t}(Y_{t+1} = y_{t+1})$$

for $A \subset \{(Y_1, \dots, Y_t) = \xi_t\}$. We denote with $\mathcal{Q}^{(\xi_t, Y_{t+1})} \mathcal{Q}^{\xi_t}$ the collection of these probability measures.

(ii) The set $\mathcal{Q}$ is called consistent *if for all $t \in \{0, 1, \dots, T-1\}$ and $\xi_t \in \mathcal{H}_t$ it holds*

$$\mathcal{Q}^{\xi_t} = \mathcal{Q}^{(\xi_t, Y_{t+1})} \mathcal{Q}^{\xi_t}.$$

By the formula of total probability, the inclusion $\mathcal{Q}^{\xi_t} \subset \mathcal{Q}^{(\xi_t, Y_{t+1})} \mathcal{Q}^{\xi_t}$ is always trivial.

The following simple observation shows that consistency and stability are equivalent and that the reverse implication of Corollary 1 (see the proof below) in Riedel (2004) is also true.

Proposition 3.3. *Let $\mathcal{Q}$ be a set of probability measures on $(\Omega, \mathcal{F})$. Then the following three statements are equivalent:*

(i) The set $\mathcal{Q}$ is stable.

(ii) The set $\mathcal{Q}$ is consistent.

(iii) If we define for every final value, i. e. for every $\mathcal{F}_T$–measurable random variable X (recall $|\Omega| < \infty$) and $t \in \{0, 1, \dots, T-1\}$

$$S_t^X(\omega) := \rho_t(0, \dots, 0, X)(\omega) := \sup_{Q \in \mathcal{Q}} \mathbb{E}_Q\Big[-\frac{X}{(1+r)^T} \mid \mathcal{F}_t\Big](\omega), \quad \omega \in \Omega,$$

then $(S_t^X)_{t=0,1,\dots,T-1}$ is a Q–supermartingale with respect to $(\mathcal{F}_t)_{t=0,1,\dots,T-1}$ for every $Q \in \mathcal{Q}$.

Proof. We show $(ii) \Rightarrow (iii) \Rightarrow (i) \Rightarrow (ii)$. The first two parts are just direct applications of the literature.

If $\mathcal{Q}$ is consistent and $Q \in \mathcal{Q}$, then $(S_t^X)_{t=0,1,\ldots,T-1}$ is a Q-supermartingale by Corollary 1 in Riedel (2004).

Now assume that $(S_t^X)_{t=0,1,\ldots,T-1}$ is a Q-supermartingale and let σ, τ be stopping times on $\{0, 1, \ldots, T\}$ such that $\sigma \leq \tau$. Then by the Doob decomposition and the Optional Sampling Theorem (see e. g. Lemma 7.10 and Theorem 7.12 in Kallenberg (2002)), we obtain in the same way as in the proof of Theorem 7.29 in Kallenberg (2002) that

$$S_\sigma^X \geq \mathbb{E}_Q[S_\tau^X \,|\, \mathcal{F}_\sigma].$$

Hence, Theorem 5.1 in Artzner et al. (2004) yields stability of the set $\mathcal{Q}$.

Finally, let $\mathcal{Q}$ be a stable set. To obtain consistency, we have to show

$$\mathcal{Q}^{(\xi_t, Y_{t+1})}\mathcal{Q}^{\xi_t} \subset \mathcal{Q}^{\xi_t}, \quad \xi_t \in \mathcal{H}_t, t = 0, 1, \ldots, T-1.$$

So let $t \in \{0, 1, \ldots, T-1\}$ and $\xi_t \in \mathcal{H}_t$. For every y_{t+1} with $(\xi_t, y_{t+1}) \in \mathcal{H}_{t+1}$ we choose a probability measure $Q_{y_{t+1}} \in \mathcal{Q}$ and $R \in \mathcal{Q}$. By definition of consistency, we have to show that there exists $Q_0 \in \mathcal{Q}$ such that for all $A \subset \{(Y_1, \ldots, Y_t) = \xi_t\}$ it holds

$$Q_0(A \,|\, (Y_1, \ldots, Y_t) = \xi_t) = \sum_{y_{t+1}:(\xi_t, y_{t+1}) \in \mathcal{H}_{t+1}} Q_{y_{t+1}}^{(\xi_t, y_{t+1})}(A) \cdot R^{\xi_t}(Y_{t+1} = y_{t+1}). \quad (3.4)$$

Consider now the constant stopping time $\tau := t+1$ and set $Q_{y_{t+1}} := \mathbb{P}$ for $y_{t+1} \in \text{Range}(Y_{t+1})$ with $(\xi_t, y_{t+1}) \notin \mathcal{H}_{t+1}$. Then $\{Y_{t+1} = y_{t+1}\} \in \mathcal{F}_\tau$ for all $y_{t+1} \in \text{Range}(Y_{t+1})$ and $\sum_{y_{t+1} \in \text{Range}(Y_{t+1})} \{Y_{t+1} = y_{t+1}\} = \Omega$. By Proposition 2.1 in Delbaen (2006), the following density process defines an element $Q_0 \in \mathcal{Q}$:

$$Z_k^{Q_0} := \begin{cases} Z_k^R & , \quad k \leq t+1 \\ \sum_{y_{t+1} \in \text{Range}(Y_{t+1})} \mathbb{1}_{\{Y_{t+1} = y_{t+1}\}} Z_{t+1}^R \cdot \frac{Z_k^{Q_{y_{t+1}}}}{Z_{t+1}^{Q_{y_{t+1}}}} & , \quad k > t+1. \end{cases}$$

Now, we easily see that Q_0 is indeed the composite probability measure from above.

To this extend, calculate for a final value X

$$\begin{aligned}
\mathbb{E}_{Q_0}\Big[X \,|\, \mathcal{F}_t\Big] &= \mathbb{E}\left[X \cdot \frac{Z_T^{Q_0}}{Z_t^{Q_0}} \,|\, \mathcal{F}_t\right] \\
&= \mathbb{E}\left[\mathbb{E}\left[X \cdot \underbrace{\sum_{y_{t+1}\in \mathrm{Range}(Y_{t+1})} \mathbb{1}_{\{Y_{t+1}=y_{t+1}\}} \frac{Z_{t+1}^R}{Z_t^R}}_{\mathcal{F}_{t+1}\text{-measurable}} \cdot \frac{Z_T^{Q_{y_{t+1}}}}{Z_{t+1}^{Q_{y_{t+1}}}} \,|\, \mathcal{F}_{t+1}\right] \Big|\, \mathcal{F}_t\right] \\
&= \mathbb{E}\left[\frac{Z_{t+1}^R}{Z_t^R} \cdot \sum_{y_{t+1}\in \mathrm{Range}(Y_{t+1})} \mathbb{1}_{\{Y_{t+1}=y_{t+1}\}} \cdot \mathbb{E}\left[X \cdot \frac{Z_T^{Q_{y_{t+1}}}}{Z_{t+1}^{Q_{y_{t+1}}}} \,|\, \mathcal{F}_{t+1}\right] \Big|\, \mathcal{F}_t\right] \\
&= \mathbb{E}_R\left[\sum_{y_{t+1}\in \mathrm{Range}(Y_{t+1})} \mathbb{1}_{\{Y_{t+1}=y_{t+1}\}} \cdot \mathbb{E}_{Q_{y_{t+1}}}[X \,|\, \mathcal{F}_{t+1}] \Big|\, \mathcal{F}_t\right] \\
&= \sum_{y_{t+1}\in \mathrm{Range}(Y_{t+1})} \mathbb{E}_R[\mathbb{E}_{Q_{y_{t+1}}}[X \cdot \mathbb{1}_{\{Y_{t+1}=y_{t+1}\}} \,|\, \mathcal{F}_{t+1}] \,|\, \mathcal{F}_t].
\end{aligned}$$

With $A \subset \{(Y_1, \ldots, Y_t) = \xi_t\}$ and $X := \mathbb{1}_A$ we obtain

$$\begin{aligned}
&Q_0(A \,|\, (Y_1, \ldots, Y_t) = \xi_t) \\
&= \sum_{y_{t+1}\in \mathrm{Range}(Y_{t+1})} \mathbb{E}_R[Q_{y_{t+1}}(A \cap \{Y_{t+1} = y_{t+1}\} \,|\, \mathcal{F}_{t+1}) \,|\, (Y_1, \ldots, Y_t) = \xi_t] \\
&= \sum_{y_{t+1}\in \mathrm{Range}(Y_{t+1})} Q_{y_{t+1}}\big(A \,|\, (Y_1, \ldots, Y_t, Y_{t+1}) = (\xi_t, y_{t+1})\big) \times \\
&\qquad\qquad \underbrace{R(Y_{t+1} = y_{t+1} \,|\, (Y_1, \ldots, Y_t) = \xi_t)}_{=0 \text{ for } (\xi_t, y_{t+1}) \notin \mathcal{H}_{t+1}} \\
&= \sum_{y_{t+1}:(\xi_t, y_{t+1})\in \mathcal{H}_{t+1}} Q_{y_{t+1}}^{(\xi_t, y_{t+1})}(A) \cdot R^{\xi_t}(Y_{t+1} = y_{t+1}).
\end{aligned}$$

Now, (3.4) and therefore the assertion follows. □

4 A risk measure by Pflug and Ruszczyński

In the last chapter, we gave an overview on the literature about dynamic risk measures in discrete time. It became obvious that the subject has been thoroughly dealt with theoretically. On the other hand, there is still a lack of practicable dynamic risk measures for processes that take into account the development of payments over time, except possibly for the so-called dynamic CTE introduced in Hardy and Wirch (2004). To deal with this drawback we generalize a proposal by Pflug and Ruszczyński (2001) that was further investigated in Pflug and Ruszczyński (2005) and Pflug (2006). In these works, a risk measure for processes is defined, while we now introduce an intuitive dynamic version of this risk measure. It is formulated as an optimization problem, which we are able to solve in a Markovian environment. We will later discuss the advantages of this proposal.

4.1 Definition of the dynamic risk measure

Let $T \in \mathbb{N}$. The dynamic risk measure will be defined for general integrable processes in discrete time, i. e. from the space

$$\mathcal{X} := \{(I_1, \ldots, I_T) \mid I_t \in \mathcal{X}_t,\ t = 1, \ldots, T\},$$

where

$$\mathcal{X}_t := L^1(\Omega, \mathcal{F}_t, \mathbb{P}), \quad t = 0, 1, \ldots, T.$$

Let us further introduce the spaces of all predictable processes starting at time $t+1$ for $t \in \{0, 1, \ldots, T-1\}$, i. e. let

$$\mathcal{X}^{(T-t)} := \{(X_t, \ldots, X_{T-1}) \mid X_k \in \mathcal{X}_k,\ k = t, \ldots, T-1\}.$$

To define the optimization problem we need some discounting factors and reinsurance premiums. To this extend, let $c_t \in \mathbb{R}_+$, $t = 1, \ldots, T+1$, with $c_{t+1} \le c_t$, $t = 1, \ldots, T$, and $q_t \in \mathbb{R}_+$, $t = 1, \ldots, T$, with $c_t \le q_t$, $t = 1, \ldots, T$. The economic motivation of the dynamic risk measure is as in Pflug and Ruszczyński (2001):

> Consider an insurance company that faces an income I_t in every period $t \in \{1, \ldots, T\}$. At time $t-1$, a decision to consume an amount a_{t-1} has to be made. If we denote the wealth of the company at time t by W_t,

the accumulated, non–negative wealth of the company can be defined recursively via

$$W_0 \equiv 0, \quad W_t = W_{t-1}^+ + I_t - a_{t-1}, \quad t = 1, \ldots, T.$$

If $W_t < 0$, the company faces a loss at time t which is reinsured by paying an insurance premium of $q_t \cdot W_t^-$. Starting in every period $t \in \{0, 1, \ldots, T-1\}$ with the wealth $W_t \in \mathcal{X}_t$, the company faces for every sequence of decisions $(a_t, \ldots, a_{T-1}) \in \mathcal{X}^{(T-t)}$ a discounted future utility of

$$\sum_{k=t+1}^{T} \frac{1}{c_t} \cdot \left(c_k a_{k-1} - q_k W_k^-\right) + \frac{c_{T+1}}{c_t} W_T^+.$$

This motivates the definition of a dynamic risk measure as the negative supremum over all strategies of the expectation of this utility.

Definition 4.1. *Let $W_t \in \mathcal{X}_t$ be the wealth at time $t \in \{0, 1, \ldots, T-1\}$. Then we define a dynamic risk measure $\rho^{PR} = (\rho_t^{PR})_{t=0,1,\ldots,T-1}$ for $I \in \mathcal{X}$ via*

$$\rho_t^{PR}(I) := - \operatorname*{ess.\,sup}_{(a_t,\ldots,a_{T-1}) \in \mathcal{X}^{(T-t)}} \frac{1}{c_t} \cdot \mathbb{E}\left[\sum_{k=t+1}^{T} \left(c_k a_{k-1} - q_k W_k^-\right) + c_{T+1} W_T^+ \,\middle|\, \mathcal{F}_t\right].$$

Remark.

1. The dynamic risk measure ρ^{PR} is independent of the past, if the starting value of the process W_t in the definition depends only on I_t but not on $I_1, \ldots, I_{t-1}$. As we will see later, a natural choice for W_t indeed fulfills this assumption.

2. One easily sees that ρ^{PR} is monotone if $W_t \equiv 0$ holds and that the process $(\rho^{\mathrm{PR}}(I)_t)_{t=0,1,\ldots,T-1}$ is $(\mathcal{F}_t)_{t=0,1,\ldots,T-1}$–adapted. Hence, ρ^{PR} is indeed a dynamic risk measure in the sense of Definition 3.1. Further properties will be investigated in the following sections.

3. Under some additional assumptions, we will show in Section 4.3 that the optimization problem can be solved explicitly and that the essential supremum is in fact a maximum.

Before investigating further properties of the dynamic risk measure, let us first make three elementary statements.

Proposition 4.1. *Let $t \in \{0, 1, \ldots, T-1\}$ and $I \in \mathcal{X}$. Then we have the following:*

(i) It always holds

$$\sum_{k=t+1}^{T} \frac{c_k}{c_t} \mathbb{E}\left[-I_k \mid \mathcal{F}_t\right] \leq \rho_t^{PR}(I) + \frac{c_{t+1}}{c_t} W_t^+ \leq - \sum_{k=t+1}^{T} \mathbb{E}\left[\frac{c_{k+1}}{c_t} I_k^+ - \frac{q_k}{c_t} I_k^- \mid \mathcal{F}_t\right].$$

(ii) If $\mathbb{E}\, W_t^+ < \infty$ *then*

$$\mathbb{E}\left[|\rho_t^{PR}(I)|\right] < \infty.$$

(iii) If we chose $c_k = q_k = c_{T+1}$, $k = 1, \dots, T$, *it holds*

$$\rho_t^{PR}(I) = -W_t^+ - \sum_{k=t+1}^{T} \mathbb{E}\left[I_k \mid \mathcal{F}_t\right]$$

and the optimal strategy is given by $(a_t^*, \dots, a_{T-1}^*) = (W_t^+, \dots, W_{T-1}^+)$.

Proof. We only need to show part (i). The rest follows immediately from this, since the I_k, $k = 1, \dots, T$, are assumed to be integrable.

Because of the relation $q_k \geq c_k \geq c_{k+1}$, $k = 1, \dots, T$, we have for an arbitrary strategy $(a_t, \dots, a_{T-1}) \in \mathcal{X}^{(T-t)}$

$$\begin{aligned}
&\sum_{k=t+1}^{T} \left(c_k a_{k-1} - q_k W_k^-\right) + c_{T+1} W_T^+ \\
&\leq \sum_{k=t+1}^{T} c_k \left(a_{k-1} - W_k^-\right) + c_{T+1} W_T^+ = \sum_{k=t+1}^{T} c_k \left(W_k - W_k^+ + a_{k-1}\right) + c_{T+1} W_T^+ \\
&= \sum_{k=t+1}^{T} c_k \left(W_{k-1}^+ + I_k - W_k^+\right) + c_{T+1} W_T^+ \\
&= \sum_{k=t+1}^{T} c_k I_k + \sum_{k=t+1}^{T} c_k W_{k-1}^+ - \sum_{k=t+1}^{T} c_k W_k^+ + c_{T+1} W_T^+ \\
&= \sum_{k=t+1}^{T} c_k I_k + \sum_{k=t}^{T} c_{k+1} W_k^+ - \sum_{k=t+1}^{T} c_k W_k^+ \\
&\leq \sum_{k=t+1}^{T} c_k I_k + c_{t+1} W_t^+ + \sum_{k=t+1}^{T} c_k W_k^+ - \sum_{k=t+1}^{T} c_k W_k^+ = \sum_{k=t+1}^{T} c_k I_k + c_{t+1} W_t^+.
\end{aligned}$$

By this, the lower bound follows.

Consider the admissible strategy $(a_t, \dots, a_{T-1}) = (W_t^+, \dots, W_{T-1}^+) \in \mathcal{X}^{(T-t)}$. The wealth process becomes $W_k = I_k$, $k = t+1, \dots, T$, and we obtain

$$\rho_t^{\mathrm{PR}}(I) \leq -\frac{c_{t+1}}{c_t} W_t^+ - \frac{1}{c_t} \cdot \mathbb{E}\left[q_{t+1} I_{t+1}^- - \sum_{k=t+2}^{T} \left(c_k I_{k-1}^+ - q_k I_k^-\right) + c_{T+1} I_T^+ \,\middle|\, \mathcal{F}_t \right],$$

hence the upper bound. □

Remark. The lower bound provided in the first part will give us in some special cases a simple way to obtain an optimal strategy. This is possible if we can define a strategy and show that the lower bound is already attained by inserting this strategy. Part (ii) shows that the optimization problem introduced to define ρ_t^{PR} has indeed a finite solution if all relevant expectations exist. From the third part we observe that in the case of constant parameters, the solution is trivial.

4.2 Properties of the dynamic risk measure

Let us now take a closer look at the properties of the dynamic risk measure in comparison with the setting of Section 3.2. As a preparation, note that the optimization problem at time t introduced above relies on the value of the wealth W_t. By a simple transformation, it is possible to consider an optimization problem that starts with $W_t \equiv 0$.

To this extend, fix $t \in \{0, 1, \ldots, T-1\}$ and $(a_t, \ldots, a_{T-1}) \in \mathcal{X}^{(T-t)}$. For another sequence of decisions $(a'_t, \ldots, a'_{T-1}) \in \mathcal{X}^{(T-t)}$ and an income process $I \in \mathcal{X}$, we introduce

$$W'_t \equiv 0, \quad W'_k = (W'_{k-1})^+ + I_k - a'_{k-1}, \quad k = t+1, \ldots, T.$$

Here, we choose $a'_t = a_t - W_t^+$ and $a'_k = a_k$, $k = t+1, \ldots, T-1$. Consequently, $W_{t+1} = W_t^+ + I_{t+1} - a_t = W'_{t+1}$ and inductively, $W_k = W'_k$, $k = t+1, \ldots, T$. Because W_t^+ is $\mathcal{F}_t$-measurable, we have $(a'_t, \ldots, a'_{T-1}) \in \mathcal{X}^{(T-t)}$ and obtain

$$\begin{aligned}
&c_t \cdot \rho_t^{\mathrm{PR}}(I) \\
&= - \operatorname*{ess.\,sup}_{a=(a_t,\ldots,a_{T-1}) \in \mathcal{X}^{(T-t)}} \mathbb{E}\left[\sum_{k=t+1}^{T} \left(c_k a_{k-1} - q_k W_k^-\right) + c_{T+1} W_T^+ \,\middle|\, \mathcal{F}_t \right] \\
&= - \operatorname*{ess.\,sup}_{a \in \mathcal{X}^{(T-t)}} \mathbb{E}\left[c_{t+1} a_t - q_k W_{t+1}^- + \sum_{k=t+2}^{T} \left(c_k a_{k-1} - q_k W_k^-\right) + c_{T+1} W_T^+ \,\middle|\, \mathcal{F}_t \right] \\
&= - \operatorname*{ess.\,sup}_{a \in \mathcal{X}^{(T-t)}} \mathbb{E}\left[c_{t+1}(a'_t + W_t^+) - q_k W_{t+1}^- + \sum_{k=t+2}^{T} \left(c_k a_{k-1} - q_k W_k^-\right) + c_{T+1} W_T^+ \,\middle|\, \mathcal{F}_t \right] \\
&= -c_{t+1} W_t^+ - \operatorname*{ess.\,sup}_{a'=(a'_t,\ldots,a'_{T-1}) \in \mathcal{X}^{(T-t)}} \mathbb{E}\left[\sum_{k=t+1}^{T} \left(c_k a'_{k-1} - q_k (W'_k)^-\right) + c_{T+1} (W'_T)^+ \,\middle|\, \mathcal{F}_t \right].
\end{aligned}$$

The optimization problem in the last term now is indeed one which starts with $W'_t \equiv 0$. Consequently, we can write the dynamic risk measure as follows:

$$\rho_t^{\mathrm{PR}}(I) = -\frac{c_{t+1}}{c_t} W_t^+ - \operatorname*{ess.\,sup}_{a \in \mathcal{X}^{(T-t)}} \frac{1}{c_t} \cdot \mathbb{E}\left[\sum_{k=t+1}^{T} \left(c_k a_{k-1} - q_k W_k^-\right) + c_{T+1} W_T^+ \,\middle|\, \mathcal{F}_t \right],$$

where the wealth process with given $I \in \mathcal{X}$, $(a_t, \ldots, a_{T-1}) \in \mathcal{X}^{(T-t)}$ and W_t (possibly depending on $I_1, \ldots, I_t$) is defined via

$$W_{t+1} = I_{t+1} - a_t, \quad W_k = W_{k-1}^+ + I_k - a_{k-1}, \quad k = t+2, \ldots, T.$$

Proposition 4.2. *If $I \in \mathcal{X}^{(T)}$, i. e. if I is predictable, then*

$$\rho_t^{PR}(I) = -\frac{c_{t+1}}{c_t} W_t^+ - \sum_{k=t+1}^{T} \frac{c_k}{c_t} \mathbb{E}\left[I_k \mid \mathcal{F}_t\right]$$

for $t \in \{0, 1, \ldots, T-1\}$ and an optimal strategy is given by

$$(a_t^*, \ldots, a_{T-1}^*) = (I_{t+1}, \ldots, I_T) \in \mathcal{X}^{(T-t)}.$$

Proof. Using the proposed strategy (which is admissible by assumption), the resulting wealth process becomes $W_{t+1}^* = \cdots = W_T^* \equiv 0$. By inserting this strategy we can observe that the lower bound given in Proposition 4.1 (i) is actually attained:

$$\begin{aligned} & -\frac{c_{t+1}}{c_t} W_t^+ - \frac{1}{c_t} \cdot \mathbb{E}\left[\sum_{k=t+1}^{T} \left(c_k a_{k-1}^* - q_k (W_k^*)^-\right) + c_{T+1} (W_T^*)^+ \,\middle|\, \mathcal{F}_t\right] \\ & = -\frac{c_{t+1}}{c_t} W_t^+ - \sum_{k=t+1}^{T} \frac{c_k}{c_t} \mathbb{E}\left[I_k \mid \mathcal{F}_t\right]. \end{aligned}$$

□

Remark. In the case $t = 0$, this is the result given in Lemma 2 in Pflug and Ruszczyński (2001).

Now, we can easily deduce two simple properties of this dynamic risk measure, namely homogeneity and translation invariance.

Proposition 4.3. *Let $t \in \{0, 1, \ldots, T-1\}$. Then we have the following.*

(i) Let $\Lambda \in L^\infty(\Omega, \mathcal{F}_t, \mathbb{P})$ with $\Lambda > 0$. If $W_t \equiv 0$, then

$$\rho_t^{PR}(\Lambda \cdot I) = \Lambda \cdot \rho_t^{PR}(I), \quad I \in \mathcal{X}.$$

(ii) Let $Y = (0, \ldots, 0, Y_{t+1}, \ldots, Y_T) \in \mathcal{X}^{(T)}$ be a predictable process such that $\sum_{k=t+1}^{T} c_k Y_k$ is $\mathcal{F}_t$-measurable. Then

$$\rho_t^{PR}(I + Y) = \rho_t^{PR}(I) - \sum_{k=t+1}^{T} \frac{c_k}{c_t} Y_k, \quad I \in \mathcal{X}.$$

Proof. (i) Let $(a_t, \ldots, a_{T-1}) \in \mathcal{X}^{(T-t)}$. Then the resulting wealth process for the income process $\Lambda \cdot I$ becomes

$$\begin{aligned} W_t^{(\Lambda)} &\equiv 0, \\ W_k^{(\Lambda)} &= (W_{k-1}^{(\Lambda)})^+ + \Lambda I_k - a_{k-1} = \Lambda \cdot \left(W_{k-1}^+ + I_k - \frac{a_{k-1}}{\Lambda}\right), \; k = t+1, \ldots, T. \end{aligned}$$

By defining $(a'_t, \ldots, a'_{T-1}) := \Lambda^{-1}(a_t, \ldots, a_{T-1}) \in \mathcal{X}^{(T-t)}$ we have a one-to-one correspondence between the two strategies and consequently obtain

$$\begin{aligned}
&\rho_t^{\mathrm{PR}}(\Lambda \cdot I) \\
&= - \operatorname*{ess.\,sup}_{(a_t,\ldots,a_{T-1}) \in \mathcal{X}^{(T-t)}} \frac{1}{c_t} \cdot \mathbb{E}\left[\sum_{k=t+1}^{T} \left(c_k a_{k-1} - q_k (W_k^{(\Lambda)})^- \right) + c_{T+1} (W_T^{(\Lambda)})^+ \,\middle|\, \mathcal{F}_t \right] \\
&= - \operatorname*{ess.\,sup}_{(a'_t,\ldots,a'_{T-1}) \in \mathcal{X}^{(T-t)}} \frac{1}{c_t} \cdot \mathbb{E}\left[\sum_{k=t+1}^{T} \Lambda \left(c_k a'_{k-1} - q_k (W'_k)^- \right) + c_{T+1} \Lambda (W'_T)^+ \,\middle|\, \mathcal{F}_t \right] \\
&= -\Lambda \operatorname*{ess.\,sup}_{(a'_t,\ldots,a'_{T-1}) \in \mathcal{X}^{(T-t)}} \frac{1}{c_t} \cdot \mathbb{E}\left[\sum_{k=t+1}^{T} \left(c_k a'_{k-1} - q_k (W'_k)^- \right) + c_{T+1} (W'_T)^+ \,\middle|\, \mathcal{F}_t \right] \\
&= \Lambda \cdot \rho_t^{\mathrm{PR}}(I).
\end{aligned}$$

(ii) Let again $(a_t, \ldots, a_{T-1}) \in \mathcal{X}^{(T-t)}$. Then the resulting wealth process for the income process $I + Y$ becomes

$$\begin{aligned}
W_t^{(Y)} &= W_t, \quad W_{t+1}^{(Y)} = I_{t+1} + Y_{t+1} - a_t = I_{t+1} - (a_t - Y_{t+1}) \\
W_k^{(Y)} &= (W_{k-1}^{(Y)})^+ + I_k - (a_{k-1} - Y_k), \quad k = t+2, \ldots, T.
\end{aligned}$$

By defining

$$(a'_t, \ldots, a'_{T-1}) := (a_t - Y_{t+1}, \ldots, a_{T-1} - Y_T) \in \mathcal{X}^{(T-t)}$$

we obtain

$$\begin{aligned}
&\rho_t^{\mathrm{PR}}(I + Y) + \frac{c_{t+1}}{c_t} W_t^+ \\
&= - \operatorname*{ess.\,sup}_{a \in \mathcal{X}^{(T-t)}} \frac{1}{c_t} \cdot \mathbb{E}\left[\sum_{k=t+1}^{T} \left(c_k a_{k-1} - q_k (W_k^{(Y)})^- \right) + c_{T+1} (W_T^{(Y)})^+ \,\middle|\, \mathcal{F}_t \right] \\
&= - \operatorname*{ess.\,sup}_{a \in \mathcal{X}^{(T-t)}} \frac{1}{c_t} \cdot \mathbb{E}\left[\sum_{k=t+1}^{T} \left(c_k (a'_{k-1} + Y_k) - q_k (W'_k)^- \right) + c_{T+1} (W'_T)^+ \,\middle|\, \mathcal{F}_t \right] \\
&= \rho_t^{\mathrm{PR}}(I) + \frac{c_{t+1}}{c_t} W_t^+ - \sum_{k=t+1}^{T} \frac{c_k}{c_t} Y_k,
\end{aligned}$$

hence the assertion. □

Remark. 1. For $t = 0$, part (i) has been shown in Pflug and Ruszczyński (2001). Furthermore, also part (ii) was investigated, but for the less general situation where $Y = (y_1, \ldots, y_T) \in \mathbb{R}^T$.

2. Part (i) does not hold if we allow for a general W_t. As we will later see, a natural choice is $W_t = I_t + \mathrm{V@R}_\gamma(I_t \mid \mathcal{F}_{t-1})$ for some $\gamma \in (0,1)$ and $I \in \mathcal{X}$. If we choose I such that $I_{t+1} = \cdots = I_T \equiv 0$ we obtain for $t > 0$ by Proposition 4.2

$$\begin{aligned}\rho_t^{\mathrm{PR}}(\Lambda \cdot I) &= -\frac{c_{t+1}}{c_t}(\Lambda I_t + \mathrm{V@R}_\gamma(\Lambda I_t \mid \mathcal{F}_{t-1}))^+ \\ &= -\Lambda\frac{c_{t+1}}{c_t}\Big(I_t + \frac{1}{\Lambda}\mathrm{V@R}_\gamma(\Lambda I_t \mid \mathcal{F}_{t-1})\Big)^+,\end{aligned}$$

which is in general not equal to $\Lambda \cdot \rho_t^{\mathrm{PR}}(I) = -\Lambda\frac{c_{t+1}}{c_t}\Big(I_t + \mathrm{V@R}_\gamma(I_t \mid \mathcal{F}_{t-1})\Big)^+$ (only if $\Lambda \in \mathcal{X}_{t-1}$).

3. Part (ii) is similar to the translation invariance property (TI3) introduced in Section 3.2 but only allows for predictable processes.

We have seen that the dynamic risk measure ρ^{PR} is independent of the past (under a mild assumption), monotone, conditional homogeneous (if $W_t \equiv 0$, $t = 0, 1, \ldots, T-1$) and that it fulfills a certain translation invariance property. To deal with the properties of subadditivity and time–consistency we will now actually solve the optimization problem in the definition of the dynamic risk measure in a Markovian setup.

4.3 Solution via Markov decision processes

In this section, we aim to solve the optimization problem in the definition of ρ^{PR} to obtain a closed formula for this dynamic risk measure. We restrict ourselves to income processes that depend on some underlying Markov chain. For $t = 0$ and general processes, this solution is obtained also in Pflug and Ruszczyński (2005), although the methods used there are different from our approach, namely via the dual optimization problem. In addition to solving the case $t > 0$, our method also allows for a generalization, by which we can treat models with some unknown parameter. This subject will thoroughly be dealt with in the next chapter. At the end of this section we will further show that ρ^{PR} does not fit in the context of Weber (2006). To deal with the more interesting setting described in Riedel (2004) and Burgert (2005) we need some further preparations and therefore skip this to Section 4.4.

Recall that for $t \in \{0, 1, \ldots, T\}$ and $I \in \mathcal{X}$

$$\rho_t^{\mathrm{PR}}(I) := -\operatorname*{ess.\,sup}_{(a_t,\ldots,a_{T-1}) \in \mathcal{X}^{(T-t)}} \frac{1}{c_t} \cdot \mathbb{E}\left[\sum_{k=t+1}^{T} \left(c_k a_{k-1} - q_k W_k^-\right) + c_{T+1} W_T^+ \,\Big|\, \mathcal{F}_t\right].$$

where $(W_t)_{t=0,1,\ldots,T}$ is an $(\mathcal{F}_t)_{t=0,1,\ldots,T}$–adapted process introduced in Section 4.1.

As usual, let $(\Omega, \mathcal{F}, \mathbb{P})$ be a rich enough probability space equipped with a filtration $(\mathcal{F}_t)_{t=0,1,\dots,T}$. Naturally, we take $\mathcal{F} = \mathcal{F}_T$. To solve the Markov decision problem introduced below we further specify this filtration by assuming that there exist stochastically independent real-valued (therefore $(\mathcal{F}, \mathcal{B})$-measurable) random variables $Y_1, \dots, Y_T$ such that

$$\mathcal{F}_0 := \{\emptyset, \Omega\}, \quad \mathcal{F}_t := \sigma(Y_1, \dots, Y_t), \quad t = 1, \dots, T.$$

We further assume that we are given a Markov chain $(Z_t)_{t=0,1,\dots,T}$ through

$$Z_0 \equiv c \in \mathbb{R}, \quad Z_t := g_t(Z_{t-1}, Y_t), \quad t = 1, \dots, T,$$

where $g_t : \mathbb{R}^2 \to \mathbb{R}$ are $(\mathcal{B}^2, \mathcal{B})$–measurable functions. We consider only income processes $I \in \mathcal{X}$ for which I_t depends only on Z_t (and, possibly on Z_{t-1}) for $t = 1, \dots, T$. In other words, we assume that there exist $(\mathcal{B}^2, \mathcal{B})$–measurable functions $h_t^I : \mathbb{R}^2 \to \mathbb{R}$, such that

$$I_t = h_t^I(Z_{t-1}, Y_t), \quad t = 1, \dots, T. \tag{4.1}$$

Denote the set of these income processes by $\mathcal{X}^{\mathrm{M}} \subset \mathcal{X}$.

This construction might look a little odd at first glance, but two simple examples show that this is an appropriate model for economic processes.

Example 4.1. Consider an example given by Artzner and mentioned in Pflug and Ruszczyński (2005). We set $T = 3$ and throw a coin three times. Each throw t for $t = 1, 2, 3$ is modelled by a random variable Y_t through

$$\mathbb{P}(Y_t = 1) = \theta = 1 - \mathbb{P}(Y_t = 0)$$

for some $\theta \in (0, 1)$, where 1 means heads and 0 tails. There are two possible games. In game 1, the player receives 1 €, if the last throw shows heads. In game 2, the player wins 1 €, if at least two of the three throws show heads. Therefore, we can write the two income processes as

$$I^{(1)} = (0, 0, Y_3) \quad \text{and} \quad I^{(2)} = (0, 0, 1_{\{Y_1+Y_2+Y_3 \geq 2\}}).$$

Obviously, both processes are identically distributed if and only if $\theta = \frac{1}{2}$ (we exclude the trivial cases $\theta = 0$ and $\theta = 1$), i. e. if the coin is fair. They fit into our model by defining $Z_0 \equiv 0$, and $Z_t = Z_{t-1} + Y_t$ for $t = 1, 2, 3$. Furthermore, note that $I_3^{(2)}$ does not depend on $I_2^{(2)}$ and Y_3, but rather on Z_2 and Y_3 as required in the model assumptions.

Example 4.2. In the first example, I_t depends only on Z_t. Now, we consider the standard Cox-Ross-Rubinstein-model to generate an income process, where I_t depends on Z_{t-1} and Y_t. Define the distribution of Y_t by

$$\mathbb{P}(Y_t = u) = \theta = 1 - \mathbb{P}(Y_t = d),$$

where $0 < d < 1 < u$ and $\theta \in (0, 1)$. Let the price process of an asset be given by $Z_0 \equiv 1$ and $Z_t := Z_{t-1} \cdot Y_t$ for $t = 0, 1, \ldots, T$. If a policy holder has one unit of this asset in her portfolio, her income from $t-1$ to t is given by

$$I_t = Z_t - Z_{t-1} = (Y_t - 1) \cdot Z_{t-1}, \quad t = 1, \ldots, T.$$

The random variable I_t can take negative values of course. The income in period $(t-1, t]$ can not be formulated as a function of Z_t, so we have to include Z_{t-1} in defining I_t. This is why we assume (4.1).

If the probability θ given in these examples is unknown, it can be modelled by a random variable ϑ. We will treat this case in Chapter 5.

Let us now define all the quantities needed for a standard Markov decision model and consider how to reformulate the dynamic risk measure ρ^{PR}.

- The *state space* is denoted by $S \subset \mathbb{R}^2$ and equipped with the σ algebra $\mathcal{S} := \mathcal{B}_S^2$. Let $s := (w, z) \in S$ be an element of the state space, where w, z represent realizations of the wealth process (W_t) and the generating Markov chain (Z_t), respectively.

- The *action space* is $A \subset \mathbb{R}$ equipped with $\mathcal{A} := \mathcal{B}_A$. Then $a \in A$ denotes the invested amount.

- There are no restrictions on the actions. Hence, for state $s \in S$ the space of *admissible policies* is $D(s) = A$ and the *restriction set* is $D = S \times A \subset \mathbb{R}^3$.

- The *disturbance* has values in $E \subset \mathbb{R}$ equipped with $\mathcal{E} := \mathcal{B}_E$.

- The *transition function* $T_t : D \times E \to S$ at time $t = 1, \ldots, T$ is given by

$$T_t(s, a, y) := (F_t(w, h_t^I(z, y), a), g_t(z, y)), \quad (s, a, y) = (w, z, a, y) \in D \times E.$$

- The *transition law* $Q_t : D \times \mathcal{S} \to [0, 1]$ at time $t = 1, \ldots, T$ is the conditional distribution of X_t given $(X_{t-1}, a_t) = (s, a) \in D$, formally

$$Q_t(s, a; B) := \mathbb{P}(T_t(s, a, Y_t) \in B), \quad B \in \mathcal{S}.$$

- The *one-step reward function* at time $t = 1, \ldots, T$ is a measurable mapping $r_t : D \to \mathbb{R}$.

- The *terminal reward function* is a measurable mapping $V_T : S \to \mathbb{R}$.

To derive an explicit representation of $(\rho_t)_{t=0,1,\ldots,T}$ in terms of conditional static risk measures (see Section 1.4) we use the setting described above by specifying the mentioned functions:

- The transition function becomes for $t = 1, \dots, T$

$$T_t(s, a, y) := (w^+ + h_t(z, y) - a, g_t(z, y)), \quad (s, a) \in D, \ y \in Z.$$

- The reward function becomes for $t = 0, \dots, T-1$

$$r_t(s, a) = -q_t w^- + c_{t+1} a, \quad (s, a) \in D.$$

where we set for convenience $q_0 := 0$.

- The terminal reward function becomes

$$V_T(s) = c_{T+1} w^+ - q_T w^-, \quad s = (w, z) \in S.$$

Furthermore, we need to introduce a set of admissible policies.

Definition 4.2. *For $t = 0, \dots, T-1$, the set of $(T-t)$–step* admissible Markov policies *is given by*

$$F^{T-t} := \{\pi = (f_t, \dots, f_{T-1}) \mid f_k : S \to A \ \ (\mathcal{S}, \mathcal{A})\text{-meas.}, \ k = t, \dots, T-1\}.$$

Now, we are ready to rewrite the dynamic risk measure in terms of Markov decision processes. Let $t \in \{0, 1, \dots, T-1\}$. Since $\rho_t^{\mathrm{PR}}(I)$ is $\mathcal{F}_t$-measurable, it is a function of $Y_1, \dots, Y_t$. Let $y = (y_1, \dots, y_t) \in E^t$ be a realization of $(Y_1, \dots, Y_T)$ and $\omega \in \{(Y_1, \dots, Y_t) = y\}$. Furthermore, there exists a function $h_t^{W,Z} : E^t \to S$ such that $(W_t, Z_t) = h_t^{W,Z}(Y_1, \dots, Y_t)$ and therefore $(W_t, Z_t)(\omega) = h_t^{W,Z}(y)$. For every $\pi = (f_t, \dots, f_{T-1}) \in F^{T-t}$, consider the Markov decision process defined via

$$X_t := (W_t, Z_t), \quad X_s := T_s(X_{s-1}, f_{s-1}(X_{s-1}), Y_s), \ s = t+1, \dots, T.$$

Thus, because of the Markovian structure of all occurring random variables, our dynamic risk measure becomes

$$\begin{aligned}
&\rho_t^{\mathrm{PR}}(I)(\omega) \\
&= - \sup_{(a_t, \dots, a_{T-1}) \in \mathcal{X}^{(T-t)}} \frac{1}{c_t} \cdot \mathbb{E}\left[\sum_{k=t+1}^{T} \left(c_k a_{k-1} - q_k W_k^- \right) + c_{T+1} W_T^+ \,\middle|\, (Y_1, \dots, Y_t) = y \right] \\
&= - \sup_{(a_t, \dots, a_{T-1}) \in \mathcal{X}^{(T-t)}} \frac{1}{c_t} \cdot \mathbb{E}\Bigg[c_{t+1} a_t + \sum_{k=t+1}^{T-1} \left(c_{k+1} a_k - q_k W_k^- \right) \\
&\qquad\qquad + c_{T+1} W_T^+ - q_T W_T^- \,\Bigg|\, (W_t, Z_t) = h_t^{W,Z}(y) \Bigg] \\
&= - \sup_{(a_t, \dots, a_{T-1}) \in \mathcal{X}^{(T-t)}} \frac{1}{c_t} \cdot \mathbb{E}\left[q_t W_t^- + \sum_{k=t}^{T-1} r_k(X_k, a_k) + V_T(X_T) \,\middle|\, X_t = h_t^{W,Z}(y) \right] \\
&= -\frac{q_t}{c_t} (h_t^{W,Z}(y))_1^- - \sup_{\pi \in F^{T-t}} \frac{1}{c_t} \cdot \mathbb{E}\left[\sum_{k=t}^{T-1} r_k(X_k, f_k(X_k)) + V_T(X_T) \,\middle|\, X_t = h_t^{W,Z}(y) \right].
\end{aligned}$$

By defining the classic value functions via

$$V_{t,\pi}(s) := \mathbb{E}\left[\sum_{k=t}^{T-1} r_k(X_k, f_k(X_k)) + V_T(X_T)\,\middle|\, X_t = s\right], \quad s \in S,$$

for every $\pi = (f_t, \ldots, f_{T-1}) \in F^{T-t}$ and

$$V_t(s) := \sup_{\pi \in F^{T-t}} V_{t,\pi}(s), \quad s \in S,$$

we obtain for $t \in \{0, 1, \ldots, T-1\}$

$$\rho_t^{\text{PR}}(I) = -\frac{q_t}{c_t} W_t^- - \frac{1}{c_t} V_t(W_t, Z_t). \tag{4.2}$$

It is well known how to derive an explicit expression for the value functions. The following theorem is from Hernández-Lerma and Lasserre (1996) and is valid for general Borel spaces.

Theorem 4.1. *Let $t \in \{0, 1, \ldots, T-1\}$ and $s \in S$ be fixed. If the functions $J_T(s') := V_T(s')$ and*

$$J_k(s') := \sup_{a \in A} \left\{ r_k(s', a) + \mathbb{E}\left[J_{k+1}(X_{k+1}) \,\middle|\, X_k = s',\, a_k = a \right] \right\}, \quad s' \in S, \tag{4.3}$$

defined for $k = t, \ldots, T-1$ are measurable and if the supremum is attained in $a^* = f_k^*(s')$, such that $f_k^* : S \to A$ is a $(\mathcal{S}, \mathcal{A})$ measurable function, then $\pi^* := (f_t^*, \ldots, f_{T-1}^*) \in F^{T-t}$ is an optimal policy in the sense that

$$V_t(s) = V_{t,\pi^*}(s) = J_t(s).$$

Proof. This is Theorem 3.2.1 in Hernández-Lerma and Lasserre (1996). There it is formulated and proven for the case $t = 0$. But as it can either be seen from the proof given there or by adapting the argument for larger t, the general case also holds. □

We can indeed apply this technique and obtain the following result. Before, introduce for $k \in \{1, \ldots, T\}$ the safety level

$$\gamma_k := \frac{q_k - c_k}{q_k - c_{k+1}} \in (0, 1), \tag{4.4}$$

the weight factor

$$\lambda_k := \frac{c_{k+1}}{c_k} \in (0, 1) \tag{4.5}$$

and a corresponding static law invariant coherent risk measure

$$\rho^{(k)}(X) := \lambda_k \mathbb{E}[-X] + (1 - \lambda_k)\text{AV@R}_{\gamma_k}(X), \quad X \in L^1(\Omega, \mathcal{F}, \mathbb{P}).$$

Note that $\rho^{(k)}$ fits into Theorem 1.4 by using Example 1.1 with $N = 2$, $p_1 = \lambda_k$, $p_2 = 1 - \lambda_k$, $\alpha_1 = 0$ and $\alpha_2 = \gamma_k$. Hence, the risk measure also satisfies the additional properties mentioned in the theorem.

Theorem 4.2. *Let $t \in \{0, 1, \dots, T\}$ and $(w, z) \in S$. Then we have the following:*

(i) The value function is given by

$$V_t(w, z) = c_{t+1}w^+ - q_t w^- - \sum_{k=t+1}^{T} c_k \cdot \mathbb{E}\,[\rho^{(k)}(I_k \,|\, Z_{k-1}) \,|\, Z_t = z].$$

(ii) The optimal policy $\pi^* = (f_t^*, \dots, f_{T-1}^*)$ and the optimal Markov process $(X_t^*, \dots, X_T^*)$ are given via

$$f_k^*(w', z') = (w')^+ - \mathrm{V@R}_{\gamma_k}(I_{k+1} \,|\, Z_k = z'), \quad s' = (w', z') \in S,$$

for $k = t, \dots, T-1$ and via the recursive relation

$$X_t^* = (w, z)$$

and for $k = t+1, \dots, T$

$$X_k^* = (I_k + \mathrm{V@R}_{\gamma_k}(I_k \,|\, Z_{k-1} = X_{k-1,2}^*), g_k(X_{k-1,2}^*, Y_k)).$$

Proof. The proof is by backward induction on t. The case $t = T$ is trivial. Since we will need the argument used in the case $t = T-1$, we first consider this. By the value iteration (Theorem 4.1) we have for $(w, z) \in S$

$$\begin{aligned}
&V_{T-1}(w, z) + q_{T-1}w^- \\
&= \sup_{a \in A} \left\{ q_{T-1}w^- + r_{T-1}((w, z), a) + \mathbb{E}\,[V_T(X_T) \;|\; X_{T-1} = (w, z), a_{T-1} = a] \right\} \\
&= \sup_{a \in A} \left\{ c_T a + \mathbb{E}\,[c_{T+1}X_{T,1}^+ - q_T X_{T,1}^- \;|\; X_{T-1} = (w, z), a_{T-1} = a] \right\} \\
&= \sup_{a \in A} \left\{ c_T a + \mathbb{E}\,[c_{T+1}[X_{T-1,1}^+ + h_T(Z_{T-1}, Y_T) - a]^+ \right. \\
&\qquad\qquad \left. - q_T[X_{T-1,1}^+ + h_T(Z_{T-1}, Y_T) - a]^- \;|\; X_{T-1} = (w, z), a_{T-1} = a] \right\} \\
&= \sup_{a \in A} \left\{ c_T a + \mathbb{E}\,[c_{T+1}[w^+ + h_T(z, Y_T) - a]^+ - q_T[w^+ + h_T(z, Y_T) - a]^-] \right\} \\
&= c_T w^+ + c_{T+1}\mathbb{E}\,[h_T(z, Y_T)] - (c_T - c_{T+1})\mathrm{AV@R}_{\gamma_T}(h_T(z, Y_T)) \\
&= c_T w^+ + c_{T+1}\mathbb{E}\,[I_T \,|\, Z_{T-1} = z] - (c_T - c_{T+1})\mathrm{AV@R}_{\gamma_T}(I_T \,|\, Z_{T-1} = z),
\end{aligned}$$

where we used Example 1.2 in the last but one step. We note that the supremum is attained in

$$a^* = f_{T-1}^*(w, z) = -\mathrm{V@R}_{\gamma_T}(h_T(z, Y_T) + w^+) = w^+ - \mathrm{V@R}_{\gamma_T}(I_T \,|\, Z_{T-1} = z),$$

from which we see that

$$\begin{aligned}
X_T^* &= (w^+ + h_T^I(z, Y_T) - f_{T-1}^*(w, z), g_T(z, Y_T)) \\
&= (h_T^I(z, Y_T) + \mathrm{V@R}_{\gamma_T}(I_T \,|\, Z_{T-1} = z), g_T(z, Y_T)).
\end{aligned}$$

By Lemma 1.1, V_{T-1} and f_{T-1} are indeed measurable functions on S, so that Theorem 4.1 can be applied. Hence, the assertion is true for $t = T-1$.

Now assume that the assertion is true for $t \leq T-1$. Together with the value iteration this yields for $(w,z) \in S$

$$\begin{aligned}
&V_{t-1}(w,z)\\
&= \sup_{a\in A}\left\{r_{t-1}((w,z),a) + \mathbb{E}\left[V_t(X_t) \mid X_{t-1} = (w,z), a_{t-1} = a\right]\right\}\\
&= \sup_{a\in A}\Big\{ - q_{t-1}w^- + c_t a + \mathbb{E}\left[c_{t+1}X_{t,1}^+ - q_t X_{t,1}^- \mid X_{t-1} = (w,z), a_{t-1} = a\right]\\
&\quad - \sum_{k=t+1}^{T} \mathbb{E}\left[\mathbb{E}\left[c_k \cdot \rho^{(k)}(I_k \mid Z_{k-1}) \mid Z_t\right] \mid X_{t-1} = (w,z), a_{t-1} = a\right]\Big\}\\
&= \sup_{a\in A}\left\{ - q_{t-1}w^- + c_t a + \mathbb{E}\left[c_{t+1}[w^+ + h_t(z,Y_t) - a]^+ - q_t[w^+ + h_t(z,Y_t) - a]^-\right]\right\}\\
&\quad + \sum_{k=t+1}^{T} c_k \cdot \mathbb{E}\left[\mathbb{E}\left[\rho^{(k)}(I_k \mid Z_{k-1}) \mid Z_t\right] \mid Z_{t-1} = z\right].
\end{aligned}$$

Now, the supremum can be treated analogously to the case $t = T-1$, where

$$a^* = f_{t-1}^*(w,z) = -\mathrm{V@R}_{\gamma_t}(h_t(z,Y_t) + w^+) = w^+ - \mathrm{V@R}_{\gamma_t}(I_t \mid Z_{t-1} = z)$$

and

$$\begin{aligned}
X_t^* &= (w^+ + h_t^I(z,Y_t) - f_{t-1}^*(w,z), g_t(z,Y_t))\\
&= (h_t^I(z,Y_t) + \mathrm{V@R}_{\gamma_t}(I_t \mid Z_{t-1} = z), g_t(z,Y_t)).
\end{aligned}$$

This yields already part (ii) of the theorem.

To obtain the desired structure of the sum from the above calculation note that for every $k \in \{t+1,\ldots,T\}$, the conditional static risk measure $\rho^{(k)}(I_k \mid Z_{k-1})$ is a function of Z_{k-1}. Therefore the Markov property yields

$$\begin{aligned}
\mathbb{E}\left[\mathbb{E}\left[\rho^{(k)}(I_k \mid Z_{k-1}) \mid Z_t\right] \mid Z_{t-1} = z\right] &= \mathbb{E}\left[\mathbb{E}\left[\rho^{(k)}(I_k \mid Z_{k-1}) \mid Z_t, Z_{t-1}\right] \mid Z_{t-1} = z\right]\\
&= \mathbb{E}\left[\rho^{(k)}(I_k \mid Z_{k-1}) \mid Z_{t-1} = z\right].
\end{aligned}$$

Plugging things together we get

$$\begin{aligned}
&V_{t-1}(w,z)\\
&= c_t w^+ - q_{t-1}w^- + c_{t+1}\mathbb{E}\left[I_t \mid Z_{t-1} = z\right] - (c_t - c_{t+1})\mathrm{AV@R}_{\gamma_t}(I_t \mid Z_{t-1} = z)\\
&\quad + \sum_{k=t+1}^{T} \mathbb{E}\left[\rho^{(k)}(I_k \mid Z_{k-1}) \mid Z_{t-1} = z\right]\\
&= c_t w^+ - q_{t-1}w^- + \sum_{k=t}^{T} \mathbb{E}\left[\rho^{(k)}(I_k \mid Z_{k-1}) \mid Z_{t-1} = z\right],
\end{aligned}$$

hence the assertion. □

Remark. 1. By defining the dynamic risk measure, we have assumed that we are given the sequence $(q_t)_{t=1,\dots,T}$. But because there is a one-to-one correspondence between the sequences $(q_t)_{t=1,\dots,T}$ and $(\gamma_t)_{t=1,\dots,T}$ via

$$q_t = \frac{c_t - c_{t+1}\gamma_t}{1-\gamma_t},$$

and (4.4) respectively we could rather assume that the safety levels are given.

2. For $k = t+1,\dots,T$, note that since $X^*_{k-1,2}$ does only depend on z but not on w, the same holds for $X^*_{k,1}$. We conclude that the optimal wealth process

$$(W^*_{t+1},\dots,W^*_T) = (X^*_{t+1,1},\dots,X^*_{T,1})$$

is the same for every value of w.

The dynamic risk measure can now be represented as follows.

Corollary 4.1. *For the dynamic risk measure we obtain for* $t = 0,1,\dots,T-1$

$$\rho_t^{PR}(I) = -\frac{c_{t+1}}{c_t}W_t^+ + \mathbb{E}\left[\sum_{k=t+1}^{T}\frac{c_k}{c_t}\cdot\rho^{(k)}(I_k \mid Z_{k-1})\,\middle|\, Z_t\right], \qquad I \in \mathcal{X}^M.$$

Proof. Direct consequence of the previous theorem and (4.2). □

Remark. 1. For $t=0$, this result has also been obtained in Pflug and Ruszczyński (2005), where a dual approach to the optimization problem is used. One advantage of our approach is the fact that we directly obtain the optimal values of the policy and the underlying wealth process.

Also, we were able to derive a formula for ρ_t^{PR} when $t>0$, where in Pflug and Ruszczyński (2005) mainly ρ_0^{PR} is investigated.

Furthermore, the next chapter will show how to generalize this method to obtain a dynamic risk measure for a model with incomplete information.

A drawback is though, that our method can only be applied by assuming a Markovian structure of the income process I, i. e. $I \in \mathcal{X}^{\mathrm{M}}$.

2. Since $\frac{c_k}{c_t}$ is just the discount factor from time k to t, we see that the risk measure at time t is the conditional expectation under Z_t of a discounted sum of convex mixtures of two conditional static risk measure applied to each component of the process $I \in \mathcal{X}^{\mathrm{M}}$.

Corollary 4.1 now helps to answer one of the open questions on the properties of ρ^{PR}, namely about subadditivity. If we choose $W_t \equiv 0$ at every point of time $t \in \{0,1,\dots,T-1\}$, the dynamic risk measure ρ^{PR} becomes subadditive because the same holds for the components $\rho^{(k)}$, $k=1,\dots,T$, in the static sense. On the other hand, a general starting value of the wealth process at every time t does not lead to subadditivity. This can be seen by first choosing $W_t := I_t + \mathrm{V@R}_{\gamma_t}(I_t \mid Z_{t-1})$. Take now $I^{(1)} = (\xi_1, 0, \dots, 0)$ and $I^{(2)} = (\xi_2, 0, \dots, 0)$ for some $\xi_1, \xi_2 \in L^1(\Omega, \mathcal{F}_1, \mathbb{P})$. Since

$$\rho_1^{\mathrm{PR}}(I^{(1)} + I^{(2)}) = -\frac{c_2}{c_1}[\xi_1 + \xi_2 + \mathrm{V@R}_{\gamma_1}(\xi_1 + \xi_2)]^+$$

and

$$\rho_1^{\mathrm{PR}}(I^{(i)}) = -\frac{c_2}{c_1}[\xi_i + \mathrm{V@R}_{\gamma_1}(\xi_i)]^+ \quad i = 1, 2,$$

we have to specify ξ_1 and ξ_2 such that

$$\mathbb{P}\big([\xi_1 + \xi_2 + \mathrm{V@R}_{\gamma_1}(\xi_1 + \xi_2)]^+ < [\xi_1 + \mathrm{V@R}_{\gamma_1}(\xi_1)]^+ + [\xi_2 + \mathrm{V@R}_{\gamma_1}(\xi_2)]^+\big) > 0 \quad (4.6)$$

in order to show that subadditivity does not hold. But this is possible simply by letting ξ_1 have an absolutely continuous distribution on the whole real line (i. e. $0 < F_X(x) < 1$ for all $x \in \mathbb{R}$), e. g. $\xi_1 \sim \mathcal{N}(0, 1)$, and taking $\xi_2 = -\xi_1$. In this case,

$$\mathbb{P}\big([\xi_1 + \xi_2 + \mathrm{V@R}_{\gamma_1}(\xi_1 + \xi_2)]^+ = 0\big) = 1.$$

On the other hand we have

$$[\xi_1 + \mathrm{V@R}_{\gamma_1}(\xi_1)]^+ + [\xi_2 + \mathrm{V@R}_{\gamma_1}(\xi_2)]^+ \geq [\xi_1 + \mathrm{V@R}_{\gamma_1}(\xi_1)]^+$$

and of course by assumption, $\mathbb{P}(\xi_1 + \mathrm{V@R}_{\gamma_1}(\xi_1) > 0) > 0$. Hence, Equation (4.6) follows and ρ^{PR} is not subadditive.

We conclude this section by calculating the dynamic risk measure ρ^{PR} for two examples.

Example 4.3. Recall that in the example by Artzner we have $T = 3$, independent Y_1, Y_2, Y_3 with $\mathbb{P}(Y_t = 1) = \theta = 1 - \mathbb{P}(Y_t = 0)$, $t = 1, 2, 3$ for some $\theta \in (0, 1)$ and

$$Z_0 \equiv 0, \quad Z_t = Z_{t-1} + Y_t,\ t = 1, 2, 3.$$

Consider the two income processes $I^{(1)} = (0, 0, Y_3)$ and $I^{(2)} = (0, 0, 1_{\{Y_1 + Y_2 + Y_3 \geq 2\}})$. Because of $I_t^{(i)} \equiv 0$, $i = 1, 2$, $t = 1, 2$ we set $W_t^{(i)} \equiv 0$, $i = 1, 2$, $t = 1, 2$. Corollary 4.1 now yields

$$\rho_t^{\mathrm{PR}}(I^{(i)}) = \frac{c_3}{c_t}\mathbb{E}\left[\rho^{(3)}(I_3^{(i)} \mid Z_2) \,\middle|\, Z_t\right], \quad i = 1, 2,\ t = 0, 1, 2.$$

This gives

$$\rho_0^{\mathrm{PR}}(I^{(1)}) = c_3 \cdot \rho^{(3)}(Y_3),$$
$$\rho_t^{\mathrm{PR}}(I^{(1)}) = \frac{1}{c_t}\rho_0^{\mathrm{PR}}(I^{(1)}), \quad t = 1, 2,$$

where

$$\mathbb{E}\left[-Y_3\right] = -\theta, \quad \mathrm{AV@R}_{\gamma_3}(Y_3) = -\mathbb{1}_{[0,\theta]}(\gamma_3)\frac{\theta - \gamma_3}{1 - \gamma_3}.$$

and therefore

$$\rho^{(3)}(Y_3) = -\lambda_3 \cdot \theta - (1 - \lambda_3) \cdot \mathbb{1}_{[0,\theta]}(\gamma_3)\frac{\theta - \gamma_3}{1 - \gamma_3}.$$

We see that all components of $\rho^{\mathrm{PR}}(I^{(1)}) = (\rho_t^{\mathrm{PR}}(I^{(1)}))_{t=0,1,2}$ are constant over all states $\omega \in \Omega$ and are only a discounted version of $\rho_0^{\mathrm{PR}}(I^{(1)})$. This has the natural interpretation that the gain of information over time does not change the risk assessment if the income only consists of a final value which is also stochastically independent of the past.

This does not hold for the process $I^{(2)}$ though, where additional information changes the value of the risk measure. We obtain

$$\begin{aligned}
\rho_0^{\mathrm{PR}}(I^{(2)}) &= c_3 \cdot \mathbb{E}\left[\rho^{(3)}(1_{\{Y_1+Y_2+Y_3 \geq 2\}} \,|\, Y_1 + Y_2)\right] \\
&= c_3 \cdot \big(2(1-\theta)\theta \cdot \rho^{(3)}(Y_3) + \theta^2 \cdot (-1)\big) \\
&= c_3\big(- \lambda_3 2(\theta^2 - \theta^3) + (1 - \lambda_3)2(\theta - \theta^2) \cdot \mathrm{AV@R}_{\gamma_3}(Y_3) - \theta^2\big).
\end{aligned}$$

For $\omega \in \{Y_1 = 0\}$ we have

$$\begin{aligned}
\rho_1^{\mathrm{PR}}(I^{(2)})(\omega) &= \frac{c_3}{c_1} \cdot \mathbb{E}\left[\rho^{(3)}(1_{\{Y_1+Y_2+Y_3 \geq 2\}} \,|\, Y_1 + Y_2) \,|\, Y_1 = 0\right] \\
&= \frac{c_3}{c_1}\big(\theta \cdot \rho^{(3)}(Y_3) + (1 - \theta) \cdot 0\big) \\
&= \frac{c_3}{c_1}\big(- \lambda_3\theta^2 + (1 - \lambda_3)\theta \cdot \mathrm{AV@R}_{\gamma_3}(Y_3)\big)
\end{aligned}$$

and for $\omega \in \{Y_1 = 1\}$

$$\begin{aligned}
\rho_1^{\mathrm{PR}}(I^{(2)})(\omega) &= \frac{c_3}{c_1} \cdot \mathbb{E}\left[\rho^{(3)}(1_{\{Y_1+Y_2+Y_3 \geq 2\}} \,|\, Y_1 + Y_2) \,|\, Y_1 = 1\right] \\
&= \frac{c_3}{c_1}\big(\theta \cdot (-1) + (1 - \theta) \cdot \rho^{(3)}(Y_3)\big) \\
&= \frac{c_3}{c_1}\big(- \lambda_3(\theta - \theta^2) + (1 - \lambda_3)(1 - \theta) \cdot \mathrm{AV@R}_{\gamma_3}(Y_3) - \theta\big).
\end{aligned}$$

Furthermore,

$$\rho_2^{\mathrm{PR}}(I^{(2)})(\omega) = \begin{cases} 0 & , \quad \omega \in \{Y_1 + Y_2 = 0\}, \\ \frac{c_3}{c_2} \cdot \rho^{(3)}(Y_3) & , \quad \omega \in \{Y_1 + Y_2 = 1\}, \\ -\frac{c_3}{c_2} & , \quad \omega \in \{Y_1 + Y_2 = 2\}. \end{cases}$$

We see that after two periods, in the case $Y_1 + Y_2 = 1$ the risk for both processes is the same, since the outcome now only depends on the value Y_3, being stochastically independent of the past. In the two other situations there is either a sure gain of 0 (if $Y_1 + Y_2 = 0$) or 1 (if $Y_1 + Y_2 = 2$), which yields a risk of 0 or -1 respectively times the discounting factor $\frac{c_3}{c_2}$. Direct calculations show that with $\theta \in (0,1)$

$$\rho_1^{\mathrm{PR}}(I^{(1)})(\omega) \begin{cases} < \rho_1^{\mathrm{PR}}(I^{(2)})(\omega) & , \quad \omega \in \{Y_1 = 0\}, \\ > \rho_1^{\mathrm{PR}}(I^{(2)})(\omega) & , \quad \omega \in \{Y_1 = 1\}, \end{cases}$$

$$\rho_2^{\mathrm{PR}}(I^{(1)})(\omega) \begin{cases} < \rho_2^{\mathrm{PR}}(I^{(2)})(\omega) & , \quad \omega \in \{Y_1 + Y_2 = 0\}, \\ = \rho_2^{\mathrm{PR}}(I^{(2)})(\omega) & , \quad \omega \in \{Y_1 + Y_2 = 1\}, \\ > \rho_2^{\mathrm{PR}}(I^{(2)})(\omega) & , \quad \omega \in \{Y_1 + Y_2 = 2\}. \end{cases}$$

If θ is chosen appropriately, all orderings are possible for the case $t = 0$. But in the most important situation $\theta = \frac{1}{2}$ or equivalently, when $I^{(1)}$ and $I^{(2)}$ are identically distributed, we have

$$\rho_0^{\mathrm{PR}}(I^{(1)}) > \rho_0^{\mathrm{PR}}(I^{(2)}).$$

In particular, we see that the dynamic risk measure ρ^{PR} is not law invariant. If ρ^{PR} would fit into the context of Theorem 3.2, the risk of $I^{(1)}$ and $I^{(2)}$ should be identical at time $t = 0$, since $\mathcal{L}(I^{(1)}) = \mathcal{L}(I^{(2)})$. We have just seen though that this is not the case.

Economic interpretation. We finally give some economic interpretations for these inequalities. First, let $t = 0$. If $\theta = \frac{1}{2}$, both processes have the same distribution, but $I^{(1)}$ has a higher risk. The reason for this fact can be seen by looking at the structure of the risk measure. The risk at time $t = 0$ is the expectation of the conditional static risk measure $\rho^{(3)}$ of the final value $I_3^{(i)}$, $i = 1, 2$, given the information Z_2. When calculating the risk of $I^{(1)}$, the final payment $I_3^{(1)} = Y_3$ is stochastically independent from this information (namely the random variable Z_2). Thus, the information is not used in this case. On the other hand, calculating the risk of $I^{(2)}$ uses this additional information that is generated over time in order to diminish the risk of the process at time $t = 0$.

Now, consider $t = 1$. If we had $Y_1 = 0$, then naturally, $I^{(2)}$ has a higher risk since – to gain 1 € – both the last throws have to show heads, while for $I^{(1)}$ only the last throw has to show heads. On the other hand, if $Y_1 = 1$, we win 1 € if one (or both) of the two last throws are successful, whereas $I^{(1)}$ only provides a positive gain, if the last one shows heads. Consequently, $I^{(1)}$ has a higher risk.

Finally, let $t = 2$. Obviously, if $Y_1 + Y_2 = 1$ the risk of the two processes is identical. In the other two cases, there is no risk left for the game modelled by $I^{(2)}$, and we either have a sure gain of 0 or 1, whereas there is still some uncertainty when using $I^{(1)}$. This explains the inequalities occurring in these situations.

Example 4.4. In the Cox Ross Rubinstein model we have

$$Z_0 \equiv 1, \quad Z_t = Z_{t-1} Y_t, \; t = 1, \dots, T,$$

where $\mathbb{P}(Y_t = u) = \theta = 1 - \mathbb{P}(Y_t = d)$ for $0 \leq d < 1 \leq u$ and $t = 1, \dots, T$ with some $\theta \in (0, 1)$. The income process is given by

$$I_t = Z_t - Z_{t-1}, \quad t = 1, \dots, T.$$

We easily compute for $t \in \{1, \dots, T\}$

$$\begin{aligned}
\mathbb{E}[Y_t] &= d + \theta(u - d), \\
\mathrm{V@R}_\gamma(Y_t) &= -d - \mathbb{1}_{[0,\theta]}(\gamma)(u-d), \ \gamma \in [0,1], \\
\mathrm{AV@R}_\gamma(Y_t) &= \begin{cases} -d & , \quad \gamma = 1 \\ -d - \mathbb{1}_{[0,\theta]}(\gamma)\frac{\theta-\gamma}{1-\gamma}(u-d) & , \quad \gamma \in [0,1) \end{cases}.
\end{aligned}$$

As the wealth process we naturally take the optimal process

$$W_0 \equiv 0, \quad W_t = I_t + \mathrm{V@R}_{\gamma_t}(I_t \mid Z_{t-1}), \, t = 1, \dots, T.$$

Using conditional translation invariance and homogeneity of $\rho^{(k)}$ we obtain for $k \in \{1, \dots, T\}$

$$\rho^{(k)}(I_k \mid Z_{k-1}) = \rho^{(k)}(Z_{k-1}(Y_k - 1) \mid Z_{k-1}) = Z_{k-1} \cdot \big(1 + \rho^{(k)}(Y_k)\big).$$

This yields for $t \in \{0, 1, \dots, T-1\}$

$$\begin{aligned}
&\rho_t^{\mathrm{PR}}(I) \\
&= -\frac{c_{t+1}}{c_t} W_t^+ + \mathbb{E}\left[\sum_{k=t+1}^{T} \frac{c_k}{c_t} \cdot \rho^{(k)}(I_k \mid Z_{k-1}) \,\middle|\, Z_t \right] \\
&= -\frac{c_{t+1}}{c_t}(Z_{t-1} \cdot (Y_t - 1) + \mathrm{V@R}_{\gamma_t}(Z_{t-1} \cdot (Y_t - 1) \mid Z_{t-1}))^+ \\
&\quad + \sum_{k=t+1}^{T} \frac{c_k}{c_t} \cdot \mathbb{E}[Z_{k-1} \mid Z_t]\big(1 + \rho^{(k)}(Y_k)\big) \\
&= -\frac{c_{t+1}}{c_t} Z_{t-1}(Y_t + \mathrm{V@R}_{\gamma_t}(Y_t))^+ + \sum_{k=t+1}^{T} \frac{c_k}{c_t} \cdot \mathbb{E}\left[Z_t \prod_{j=t+1}^{k-1} Y_j \,\middle|\, Z_t \right] \cdot \big(1 + \rho^{(k)}(Y_k)\big) \\
&= -\frac{c_{t+1}}{c_t} Z_{t-1}(Y_t + \mathrm{V@R}_{\gamma_t}(Y_t))^+ + Z_t \sum_{k=t+1}^{T} \frac{c_k}{c_t} \cdot \mathbb{E}[Y_1]^{k-t-1} \cdot \big(1 + \rho^{(k)}(Y_k)\big).
\end{aligned}$$

4.4 A stable representation result

Using the result of Corollary 4.1 and the Markovian structure of the occurring random variables, we can write the risk measure ρ^{PR} for $t \in \{0, 1, \dots, T-1\}$ as

$$\rho_t^{\mathrm{PR}}(I) = -\frac{c_{t+1}}{c_t} W_t^+ + \mathbb{E}\left[\sum_{k=t+1}^{T} \frac{c_k}{c_t} \cdot \rho^{(k)}(I_k \mid \mathcal{F}_{k-1}) \,\middle|\, \mathcal{F}_t \right] \tag{4.7}$$

for every $I \in \mathcal{X}^{\mathrm{M}}$. The results in the remaining part of this chapter regarding the representation (4.7) are valid for general processes $I \in \mathcal{X}$, though.

The aim of this section is to give a different representation of the dynamic risk measure in terms of stable sets of probability measures. Concluding these investigations we will give a simple counterexample to show that ρ^{PR} does not fit into the context of Theorem 3.1. This observation will already be intuitively clear from our derived stable representation.

Let us first examine for $k = 1, \ldots, T$ a certain set of probability measures, namely define

$$\mathcal{Q}_k := \Big\{ Q \in \mathcal{P}_k^* \mid \frac{c_{k+1}}{c_k} \leq L_k^Q \leq \frac{q_k}{c_k},\ \mathbb{E}\,[L_k^Q \mid \mathcal{F}_{k-1}] \equiv 1 \Big\}.$$

Similar to the notions introduced in Section 3.2, let $\mathcal{P}$ be the set of all probability measures on Ω, $\mathcal{P}_k^*$ the subset of all probability measures which are absolutely continuous with respect to $\mathbb{P}$ on $\mathcal{F}_k$ and $\mathcal{P}^e$ include all probability measures which are equivalent to $\mathbb{P}$. If $Q \in \mathcal{P}_k^*$, let L_k^Q be the resulting $\mathcal{F}_k$-measurable density.

In order to show that $\mathcal{Q}_k$ is stable, we first give a reformulation of the set.

Lemma 4.1. *For the set $\mathcal{Q}_k$ defined above it holds:*

$$\mathcal{Q}_k = \Big\{ Q \in \mathcal{P}_k^* \mid \frac{c_{k+1}}{c_k} \cdot \mathbb{P}(B) \leq Q(B) \leq \frac{q_k}{c_k} \cdot \mathbb{P}(B),\ B \in \mathcal{F}_k,\ Q = \mathbb{P} \text{ on } \mathcal{F}_{k-1} \Big\}.$$

Proof. Let $Q \in \mathcal{P}_k^*$ and $B \in \mathcal{F}_{k-1} \subset \mathcal{F}_k$. By definition, we always have

$$\int_B L_k^Q \,\mathrm{d}\mathbb{P} = Q(B).$$

Combining this with $\mathbb{P}(B) = \int_B 1 \,\mathrm{d}\mathbb{P}$ we conclude that

$$\mathbb{E}\,[L_k^Q \mid \mathcal{F}_{k-1}] \equiv 1 \iff Q(B) = \mathbb{P}(B),\ \forall\, B \in \mathcal{F}_{k-1}.$$

The second properties are also equivalent, which can be seen as follows:

$$\begin{aligned}
& \frac{c_{k+1}}{c_k} \cdot \mathbb{P}(B) \leq Q(B) \leq \frac{q_k}{c_k} \cdot \mathbb{P}(B),\ \forall\, B \in \mathcal{F}_k \\
\iff\ & \frac{c_{k+1}}{c_k} \cdot \int_B \mathrm{d}\mathbb{P} \leq \int_B \mathrm{d}Q \leq \frac{q_k}{c_k} \cdot \int_B \mathrm{d}\mathbb{P},\ \forall\, B \in \mathcal{F}_k \\
\iff\ & \int_B \frac{c_{k+1}}{c_k} \,\mathrm{d}\mathbb{P} \leq \int_B L_k^Q \,\mathrm{d}\mathbb{P} \leq \int_B \frac{q_k}{c_k} \,\mathrm{d}\mathbb{P},\ \forall\, B \in \mathcal{F}_k \\
\iff\ & \frac{c_{k+1}}{c_k} \leq L_k^Q \leq \frac{q_k}{c_k}.
\end{aligned}$$

The last equivalence follows since L_k^Q is $\mathcal{F}_k$-measurable. □

Remark. By the relationship

$$\frac{c_{k+1}}{c_k} \cdot \mathbb{P}(B) \leq Q(B) \leq \frac{q_k}{c_k} \cdot \mathbb{P}(B),\ \forall\, B \in \mathcal{F}_k,$$

for some $Q \in \mathcal{Q}_k$ we see that on $\mathcal{F}_k$ not only $Q \ll \mathbb{P}$ but also $\mathbb{P} \ll Q$ and therefore $Q \sim \mathbb{P}$ holds.

From this result we can straightforwardly deduce that $\mathcal{Q}_k$ is stable. This is interesting in itself such that we formulate this property as a proposition.

Proposition 4.4. *The set of probability measures $\mathcal{Q}_k$ is stable on $\mathcal{F}_k$.*

Proof. Recall that we have to show for any pair $Q_1, Q_2 \in \mathcal{Q}_k$ and any stopping time $\tau \leq k$ that the probability measure Q^τ defined via its $\mathbb{P}$-density on $\mathcal{F}_k$

$$L_k^{Q^\tau} := L_\tau^{Q_1} \frac{L_k^{Q_2}}{L_\tau^{Q_2}}$$

is again in $\mathcal{Q}_k$.

For any $Q \in \mathcal{Q}_k$, we have $\mathbb{E}\,[L_k^Q \,|\, \mathcal{F}_{k-1}] \equiv 1$ and therefore also

$$L_t^Q = \mathbb{E}\,[L_k^Q \,|\, \mathcal{F}_t] \equiv 1, \quad t = 0, 1, \ldots, k-1. \tag{4.8}$$

We immediately obtain

$$L_k^{Q^\tau}(\omega) = \begin{cases} L_k^{Q_1}(\omega) \frac{L_k^{Q_2}(\omega)}{L_k^{Q_2}(\omega)} = L_k^{Q_1}(\omega) & , \quad \omega \in \{\tau = k\} \\ L_t^{Q_1}(\omega) \frac{L_k^{Q_2}(\omega)}{L_t^{Q_2}(\omega)} \overset{(4.8)}{=} L_k^{Q_2}(\omega) & , \quad \omega \in \{\tau = t\},\ t = 0, 1, \ldots, k-1. \end{cases}$$

From this we get for $B \in \mathcal{F}_k$

$$\begin{aligned} Q^\tau(B) &= \mathbb{E}\,[\mathbb{1}_B L_k^{Q^\tau}] = \mathbb{E}\,[\mathbb{1}_{B \cap \{\tau = k\}} L_k^{Q_1}] + \mathbb{E}\,[\mathbb{1}_{B \cap \{\tau < k\}} L_k^{Q_2}] \\ &= Q_1(B \cap \{\tau = k\}) + Q_2(B \cap \{\tau < k\}). \end{aligned} \tag{4.9}$$

Now, stability follows easily:

- First, let $B \in \mathcal{F}_{k-1}$. Because $\tau \leq k$, we have $\{\tau = k\} \in \mathcal{F}_{k-1}$ and therefore also $B \cap \{\tau < k\}$, $B \cap \{\tau = k\} \in \mathcal{F}_{k-1}$. Consequently, since $Q_1, Q_2 \in \mathcal{Q}_k$,

$$\begin{aligned} Q^\tau(B) &= Q_1(B \cap \{\tau = k\}) + Q_2(B \cap \{\tau < k\}) \\ &= \mathbb{P}(B \cap \{\tau = k\}) + \mathbb{P}(B \cap \{\tau < k\}) = \mathbb{P}(B). \end{aligned}$$

- Now, let $B \in \mathcal{F}_k$. We obtain $B \cap \{\tau < k\}$, $B \cap \{\tau = k\} \in \mathcal{F}_k$ and obtain again with $Q_1, Q_2 \in \mathcal{Q}_k$

$$\begin{aligned} Q^\tau(B) &= Q_1(B \cap \{\tau = k\}) + Q_2(B \cap \{\tau < k\}) \\ &\leq \frac{q_k}{c_k} \cdot \mathbb{P}(B \cap \{\tau = k\}) + \frac{q_k}{c_k} \cdot \mathbb{P}(B \cap \{\tau < k\}) \\ &= \frac{q_k}{c_k} \cdot \mathbb{P}(B). \end{aligned}$$

In the same manner, $Q^\tau(B) \geq \frac{c_{k+1}}{c_k} \cdot \mathbb{P}(B)$ holds and hence the assertion is proved. □

Using this lemma, we can now reformulate the components of our dynamic risk measure.

Lemma 4.2. *Let $k \in \{1, \ldots, T\}$. Then:*

(i) We have

$$\rho^{(k)}(X \mid \mathcal{F}_{k-1}) = \underset{Q \in \mathcal{Q}_k}{\text{ess.}\sup} \mathbb{E}_Q[-X \mid \mathcal{F}_{k-1}], \quad X \in L^1(\Omega, \mathcal{F}_k, \mathbb{P}).$$

(ii) For $t = 0, \ldots, k-1$, it holds

$$\mathbb{E}\left[\rho^{(k)}(X \mid \mathcal{F}_{k-1}) \mid \mathcal{F}_t\right] = \underset{Q \in \mathcal{Q}_k}{\text{ess.}\sup} \mathbb{E}_Q[-X \mid \mathcal{F}_t], \quad X \in L^1(\Omega, \mathcal{F}_k, \mathbb{P}).$$

Proof. By definition of the Average Value at Risk we initially get for arbitrary $X \in L^1(\Omega, \mathcal{F}_k, \mathbb{P})$

$$\begin{aligned}
&\rho^{(k)}(X \mid \mathcal{F}_{k-1}) \\
&= \lambda_k \cdot \mathbb{E}\left[-X \mid \mathcal{F}_{k-1}\right] + (1 - \lambda_k) \cdot \text{AV@R}_{\gamma_k}(X \mid \mathcal{F}_{k-1}) \\
&= \lambda_k \cdot \mathbb{E}\left[-X \mid \mathcal{F}_{k-1}\right] + (1 - \lambda_k) \cdot \underset{Y\ \mathcal{F}_{k-1}\text{-meas.}}{\text{ess.}\inf} \left\{Y + \frac{1}{1-\gamma_k} \cdot \mathbb{E}\left[(-X-Y)^+ \mid \mathcal{F}_{k-1}\right]\right\} \\
&= \text{ess.}\sup\{\mathbb{E}\left[-X \cdot L \mid \mathcal{F}_{k-1}\right] \mid L\ \mathcal{F}_k\text{-meas.}, \lambda_k \leq L \leq 1 + \varepsilon, \mathbb{E}\left[L \mid \mathcal{F}_{k-1}\right] \equiv 1\},
\end{aligned}$$

where the last equality in the above calculation follows from equality (6.14) in Ruszczyński and Shapiro (2006) and where $\gamma_k = \frac{\varepsilon}{1-\lambda_k+\varepsilon}$ which is equivalent to

$$\varepsilon = \frac{q_k}{c_k} - 1.$$

Let us now identify each occurring L with a probability measure such that $Q \ll \mathbb{P}$ on $\mathcal{F}_k$ by setting $L_k^Q := L$. We obtain

$$\begin{aligned}
\rho^{(k)}(X \mid \mathcal{F}_{k-1}) &= \underset{Q \in \mathcal{Q}_k}{\text{ess.}\sup} \mathbb{E}\left[-X \cdot L_k^Q \mid \mathcal{F}_{k-1}\right] \overset{(3.2)}{=} \underset{Q \in \mathcal{Q}_k}{\text{ess.}\sup} \mathbb{E}_Q[-X \cdot L_{k-1}^Q \mid \mathcal{F}_{k-1}] \\
&= \underset{Q \in \mathcal{Q}_k}{\text{ess.}\sup} \mathbb{E}_Q[-X \mid \mathcal{F}_{k-1}],
\end{aligned}$$

since $L_{k-1}^Q = \mathbb{E}\left[L_k^Q \mid \mathcal{F}_{k-1}\right] \equiv 1$ for every $Q \in \mathcal{Q}_k$.

For every $t < k$ we also have $L_t^Q = \mathbb{E}\left[L_k^Q \mid \mathcal{F}_t\right] \equiv 1$ for every $Q \in \mathcal{Q}_k$. Furthermore, the remark after Lemma 4.1 yields $\mathcal{Q}_k = \mathcal{Q}_k^e$. We consequently obtain by Proposition 3.2

$$\begin{aligned}
&\mathbb{E}\left[\rho^{(k)}(X \mid \mathcal{F}_{k-1}) \mid \mathcal{F}_t\right] \\
&= \mathbb{E}\left[\underset{Q \in \mathcal{Q}_k}{\text{ess.}\sup} \mathbb{E}_Q[-X \mid \mathcal{F}_{k-1}] \mid \mathcal{F}_t\right] = \underset{Q \in \mathcal{Q}_k}{\text{ess.}\sup} \mathbb{E}\left[\mathbb{E}_Q[-X \mid \mathcal{F}_{k-1}] \mid \mathcal{F}_t\right] \\
&= \underset{Q \in \mathcal{Q}_k}{\text{ess.}\sup} \mathbb{E}\left[\underbrace{L_{k-1}^Q}_{\equiv 1} \cdot \mathbb{E}_Q[-X \mid \mathcal{F}_{k-1}] \mid \mathcal{F}_t\right] \overset{(3.2)}{=} \underset{Q \in \mathcal{Q}_k}{\text{ess.}\sup} \mathbb{E}_Q\left[\underbrace{L_t^Q}_{\equiv 1} \cdot \mathbb{E}_Q[-X \mid \mathcal{F}_{k-1}] \mid \mathcal{F}_t\right] \\
&= \underset{Q \in \mathcal{Q}_k}{\text{ess.}\sup} \mathbb{E}_Q[-X \mid \mathcal{F}_t],
\end{aligned}$$

hence the assertion. □

Combining Lemma 4.2 and (4.7) immediately yields the main result of this subsection.

Theorem 4.3. *For every $t \in \{0, 1, \dots, T-1\}$ it holds*

$$\rho_t^{PR}(I) = -\frac{c_{t+1}}{c_t} W_t^+ + \sum_{k=t+1}^{T} \frac{c_k}{c_t} \cdot \operatorname{ess.\,sup}_{Q \in \mathcal{Q}_k} \mathbb{E}_Q[-I_k \mid \mathcal{F}_t], \quad I \in \mathcal{X}.$$

For $k = 1, \dots, T$, the set $\mathcal{Q}_k$ is stable on $\mathcal{F}_k$. Furthermore, it does not depend on t.

We now investigate the relationship of this representation with the literature. As a first note, we can observe that the risk measure does in general not fit into the setting of Theorem 3.1. This is already intuitively clear, since in the cited theorem, the class of dynamic risk measures depends at time t merely on the discounted sum of the future payments, while our risk measure seems to be able to assign different values to different processes where this discounted sum might be the same. Take the natural choice of discounting factors

$$c_t := (1+r)^{-t}, t = 0, 1, \dots, T+1,$$

for some $r > -1$ and assume that there exists some stable set $\mathcal{Q}$ such that

$$\rho_t^{\mathrm{PR}}(I) = \operatorname{ess.\,sup}_{Q \in \mathcal{Q}} \mathbb{E}_Q\Big[-\sum_{k=t}^{T} \frac{c_k}{c_t} I_k \,\Big|\, \mathcal{F}_t\Big], \quad I \in \mathcal{X}. \tag{4.10}$$

By Theorem 3.1, the dynamic risk measure fulfills the time-consistency property, i. e. for all $I \in \mathcal{X}$, stopping times $\sigma \leq \tau$ and $Z \in L^1(\Omega, \mathcal{F}_\tau, \mathbb{P})$ holds

$$\rho_\sigma^{\mathrm{PR}}(I + Z \cdot e_\tau) = \rho_\sigma^{\mathrm{PR}}(I + (1+r)^{T-\tau} \cdot Z \cdot e_T).$$

A most simple counterexample now shows that this can indeed not be the case. Take $T = 2$, $\sigma \equiv 0$, $\tau \equiv 1$ and let Z some integrable $\mathcal{F}_\tau = \mathcal{F}_1$-measurable random variable. We obtain for the process $I = (I_1, I_2) = (0, 0)$

$$\begin{aligned}\rho_\sigma^{\mathrm{PR}}(I + Z \cdot e_\tau) &= \rho_0^{\mathrm{PR}}(Z, 0) = c_1 \cdot \rho^{(1)}(Z) \\ &= c_1 \cdot (\lambda_1 \mathbb{E}[-Z] + (1-\lambda_1) \mathrm{AV@R}_{\gamma_1}(Z))\end{aligned}$$

and

$$\begin{aligned}\rho_\sigma^{\mathrm{PR}}(I + (1+r)^{T-\tau} \cdot Z \cdot e_T) &= \rho_0^{\mathrm{PR}}\Big(0, \frac{c_1}{c_2} \cdot Z\Big) = c_2 \cdot \mathbb{E}\Big[\rho^{(2)}\Big(\frac{c_1}{c_2} \cdot Z \,\Big|\, \mathcal{F}_1\Big)\Big] \\ &= c_2 \cdot \mathbb{E}\Big[-\frac{c_1}{c_2} \cdot Z\Big] = c_1 \cdot \mathbb{E}[-Z].\end{aligned}$$

Consequently, equality of the last two terms would hold if and only if

$$\mathbb{E}[-Z] = \mathrm{AV@R}_{\gamma_1}(Z),$$

which in general is only fulfilled if $\gamma_1 = 0$ or if Z is a constant. We conclude that a representation as in (4.10) is not possible for our dynamic risk measure ρ^{PR}.

But since the set $\mathcal{Q}_T$ is stable (on $\mathcal{F}_T$) we can use it to construct a new dynamic risk measure in the setting of Theorem 3.1 via

$$\rho_t^{\text{M}}(I) := \operatorname*{ess.\,sup}_{Q \in \mathcal{Q}_T} \mathbb{E}_Q\Big[-\sum_{k=t}^{T} \frac{c_k}{c_t} I_k \,|\, \mathcal{F}_t \Big] = \mathbb{E}\Big[\rho^{(T)}\Big(\sum_{k=t}^{T} \frac{c_k}{c_t} I_k \,|\, \mathcal{F}_{T-1} \Big) \,|\, \mathcal{F}_t \Big], \quad I \in \mathcal{X}.$$

Secondly, we now compare our stable sets with some other sets from the literature. More precisely, in Artzner et al. (2004), a dynamic version of Average Value–at–Risk $\rho^{\text{A}} = (\rho_t^{\text{A}})_{t=0,1,\ldots,T-1}$ for final payments is proposed via

$$\rho_t^{\text{A}}(0,\ldots,0,X) := \text{AV@R}_\gamma^{(t,T)}(X) := \operatorname*{ess.\,sup}_{Q \in \mathcal{Q}'_t} \mathbb{E}\,[-X \cdot L_T^Q \,|\, \mathcal{F}_t], \tag{4.11}$$

for $X \in L^1(\Omega, \mathcal{F}_T, \mathbb{P})$ and $\gamma \in (0,1)$, where

$$\mathcal{Q}'_t := \Big\{ Q \in \mathcal{P}_T^* \,|\, 0 \le L_T^Q \le \frac{1}{1-\gamma},\ \mathbb{E}\,[L_T^Q \,|\, \mathcal{F}_t] \equiv 1 \}, \quad t = 0,1,\ldots,T-1.$$

Writing down our proposal for a dynamic risk measure for final payments, we easily see by arguments used above that

$$\begin{aligned}
\rho_t^{\text{PR}}(0,\ldots,0,X) &= \operatorname*{ess.\,sup}_{Q \in \mathcal{Q}_T} \mathbb{E}_Q[-X \,|\, \mathcal{F}_t] = \mathbb{E}\,[\rho^{(T)}(X \,|\, \mathcal{F}_{T-1}) \,|\, \mathcal{F}_t] \\
&= \lambda_T \cdot \mathbb{E}\,[-X \,|\, \mathcal{F}_t] + (1-\lambda_T) \cdot \mathbb{E}\,[\text{AV@R}_{\gamma_T}(X \,|\, \mathcal{F}_{T-1}) \,|\, \mathcal{F}_t] \\
&= \lambda_T \cdot \mathbb{E}\,[-X \,|\, \mathcal{F}_t] + (1-\lambda_T) \cdot \operatorname*{ess.\,sup}_{Q \in \mathcal{Q}^F} \mathbb{E}_Q[-X \,|\, \mathcal{F}_t],
\end{aligned}$$

where

$$\mathcal{Q}^F := \Big\{ Q \in \mathcal{P}_T^* \,|\, 0 \le L_T^Q \le \frac{1}{1-\gamma_T},\ \mathbb{E}\,[L_T^Q \,|\, \mathcal{F}_{T-1}] \equiv 1 \Big\}.$$

Stability of the set $\mathcal{Q}^F$ can be proved analogously to Proposition 4.4. In order to compare the two dynamic risk measures for final payments take for now $\lambda_T = 0$. We obtain

$$\rho_t^{\text{PR}}(0,\ldots,0,X) = \operatorname*{ess.\,sup}_{Q \in \mathcal{Q}^F} \mathbb{E}\,[-X \cdot L_T^Q \,|\, \mathcal{F}_t].$$

in contrast to (4.11).

Despite the similarity with the stable set $\mathcal{Q}^F = \mathcal{Q}'_{T-1}$, the remaining sets $\mathcal{Q}'_t$, $t = 0,1,\ldots,T-2$ are not stable if $T > 1$ (see Artzner et al. (2004), Section 5.3). To see why one can not proof stability of $\mathcal{Q}'_0$, for example, in a similar way than stability of $\mathcal{Q}^F$, note that in the case $T = 2$, any stopping time τ can attain the three values $0, 1$ and 2. Consequently, for every $Q \in \mathcal{Q}'_0$ we only have $L_0^Q \equiv 1$ whereas for every $Q \in \mathcal{Q}_2$, we get $L_0^Q = L_1^Q \equiv 1$. This property is an essential step in the proof of Proposition 4.4 when determining Equation (4.9).

Furthermore, for every ρ_t^{A}, $t = 0, 1, \ldots, T-1$, a different set $\mathcal{Q}_t'$ is used. This is in contrast to the theory in Burgert (2005), where the essence of the construction of consistent dynamic risk measures is the fact that one only needs *one* stable set for every dynamic risk measure $\rho = (\rho_t)_{t=0,1,\ldots,T-1}$. These two reasons complement the given counterexample and illustrate why the dynamic risk measure ρ^{A} does not yield a time–consistent dynamic risk measure for final payments via a stable representation whereas ρ^{PR} does.

Before turning to some simple martingale properties, we are now ready to give a short summary on the properties of the dynamic risk measure ρ^{PR}. We have seen that it is in general translation invariant and monotone. Further properties can only be derived by making additional assumptions on the starting value of the wealth process at every time $t \in \{0, 1, \ldots, T-1\}$. If this W_t does only depend on I_t (but not on $I_1, \ldots, I_{t-1}$), then we also have independence of the past. This is fulfilled if we choose

$$W_t := I_t + \mathrm{V@R}_{\gamma_t}(I_t \,|\, \mathcal{F}_t), \quad t = 1, \ldots, T-1. \tag{4.12}$$

Furthermore, the properties of homogeneity and subadditivity only hold in general if $W_t \equiv 0$, $t = 1, \ldots, T-1$. It is also not possible to get time–consistency for the dynamic risk measure as long as there are not only final payments. The main conclusion of these investigations is that ρ^{PR} does not fit into the context of Theorems 3.1 and 3.2.

4.5 Martingales & co.

To conclude this chapter, we give some brief and simple statements on martingale properties of the representation (4.7). To motivate the investigations and to keep things simple, we start with the special case of a final payment and then go on to more general cases. We always consider the natural choice of the wealth process, i. e. assume (4.12).

Proposition 4.5. *Let $\xi \in L^1(\Omega, \mathcal{F}_T, \mathbb{P})$. Then $(S_t)_{t=0,1,\ldots,T-1}$ with*

$$S_t := c_t \rho_t^{\mathrm{PR}}(0, \ldots, 0, \xi), \quad t = 0, 1, \ldots, T-1,$$

is a martingale with respect to $(\mathcal{F}_t)_{t=0,1,\ldots,T-1}$.

Proof. By (4.12) we get $W_t \equiv 0$, $t = 0, 1, \ldots, T-1$. Defining $\eta := c_T \rho^{(T)}(\xi \,|\, \mathcal{F}_{T-1})$ yields

$$S_t = c_t \cdot \frac{1}{c_t} \mathbb{E}\left[c_T \rho^{(T)}(\xi \,|\, \mathcal{F}_{T-1}) \,|\, \mathcal{F}_t\right] = \mathbb{E}\left[\eta \,|\, \mathcal{F}_t\right]$$

and therefore the assertion. □

Now we treat the dynamic risk measure that was proposed in the previous section where we constructed a time-consistent variant of ρ^{PR}. Recall that we defined

$$\rho_t^{\mathrm{M}}(I) = \operatorname*{ess.\,sup}_{Q \in \mathcal{Q}_T} \mathbb{E}_Q\Big[-\sum_{k=t}^{T} \frac{c_k}{c_t} I_k \,|\, \mathcal{F}_t\Big], \quad I \in \mathcal{X},$$

where for convenience $I_0 = 0$.

Proposition 4.6. *Let $I \in \mathcal{X}$. Then $(S_t)_{t=0,1,\ldots,T-1}$ with*

$$S_t := c_t \rho_t^{\mathrm{M}}(I) - \sum_{k=0}^{t-1} c_k I_k, \quad t = 0, 1, \ldots, T-1,$$

is a martingale with respect to $(\mathcal{F}_t)_{t=0,1,\ldots,T-1}$.

Proof. We first note that by homogeneity

$$\rho_t^{\mathrm{M}}(I) = \frac{c_t}{c_T} \cdot \rho_t^{\mathrm{PR}}\Big(0, \ldots, 0, \sum_{k=t}^{T} \frac{c_k}{c_t} I_k\Big) = \rho_t^{\mathrm{PR}}\Big(0, \ldots, 0, \sum_{k=t}^{T} \frac{c_k}{c_T} I_k\Big).$$

Hence, translation invariance of ρ^{PR} (compare Proposition 4.3) yields

$$\begin{aligned} S_t &= c_t \cdot \Big(\rho_t^{\mathrm{M}}(I) - \frac{1}{c_t}\sum_{k=0}^{t-1} c_k I_k\Big) = c_t \cdot \Big(\rho_t^{\mathrm{PR}}\Big(0, \ldots, 0, \sum_{k=t}^{T} \frac{c_k}{c_T} I_k\Big) - \frac{c_T}{c_t}\sum_{k=0}^{t-1} \frac{c_k}{c_T} I_k\Big) \\ &= c_t \cdot \rho_t^{\mathrm{PR}}\Big(0, \ldots, 0, \sum_{k=0}^{T} \frac{c_k}{c_T} I_k\Big). \end{aligned}$$

With $\xi := \sum_{k=0}^{T} \frac{c_k}{c_T} I_k$ we can apply Proposition 4.5 and therefore obtain the assertion. □

Remark. Indeed, defining $(S_t)_{t=0,1,\ldots,T-1}$ as in the previous proposition by using an arbitrary dynamic risk measure in the sense of Theorem 3.1 instead of ρ^{M}, the process $(S_t)_{t=0,1,\ldots,T-1}$ is in general only a supermartingale. Also compare Remark 3.28 in Burgert (2005), where the process $\big(\frac{S_t}{c_t}\big)_{t=0,1,\ldots,T-1}$ is wrongly claimed to be a supermartingale, and the correct result in Corollary 1 of Riedel (2004) and its connection with the notion of stability investigated in Proposition 3.3.

Let us consider now how to transform ρ^{PR} into a martingale in the general case.

Proposition 4.7. *Let $I \in \mathcal{X}$. Then $(S_t)_{t=0,1,\ldots,T-1}$ with*

$$S_t := c_t \rho_t^{\mathrm{PR}}(I) + c_{t+1} W_t^{+} + \sum_{k=1}^{t} c_k \cdot \rho^{(k)}(I_k \,|\, \mathcal{F}_{k-1}), \quad t = 0, 1, \ldots, T-1,$$

is a martingale with respect to $(\mathcal{F}_t)_{t=0,1,\ldots,T-1}$.

Proof. Following the same principle as in the two previous proofs, we obtain

$$S_t \overset{(4.7)}{=} \mathbb{E}\left[\sum_{k=t+1}^{T} c_k \cdot \rho^{(k)}(I_k \,|\, \mathcal{F}_{k-1}) \,\Big|\, \mathcal{F}_t\right] + \underbrace{\sum_{k=1}^{t} c_k \cdot \rho^{(k)}(I_k \,|\, \mathcal{F}_{k-1})}_{\mathcal{F}_t\text{-measurable}}$$

$$= \mathbb{E}\left[\sum_{k=1}^{T} c_k \cdot \rho^{(k)}(I_k \,|\, \mathcal{F}_{k-1}) \,\Big|\, \mathcal{F}_t\right] = \mathbb{E}\left[\eta \,|\, \mathcal{F}_t\right]$$

with $\eta := \sum_{k=1}^{T} c_k \cdot \rho^{(k)}(I_k \,|\, \mathcal{F}_{k-1})$. Hence, the assertion follows. □

Remark. Obviously, Proposition 4.5 follows immediately from Proposition 4.7.

Economic interpretation. The interpretation of Proposition 4.5 is quite obvious. The result shows, that the process of the discounted risk is a martingale if we only have a final payment. This means that the average of the risk that we assign to the final payment (conditioned on the current available information) in the next period is the same as the risk of today. This is quite natural, since we do not receive any payments over time until the last period and therefore have no need to change the assigned risk (on average) from one period to another.

5 A Bayesian control approach

In this chapter, we generalize the definition of the dynamic risk measure from Chapter 4 for models with an unknown parameter. This is motivated by the economically interpretable properties which we derived for the standard risk measure in the previous chapter and by the solution of the introduced optimization problem via Markov decision processes.

5.1 The model

Let $(\Omega, \mathcal{F}, \mathbb{P})$ be a rich enough probability space and $Y_1, \ldots, Y_T$ real-valued (therefore $(\mathcal{F}, \mathcal{B})$-measurable) random variables on it, where $T \in \mathbb{N}$. As usual, $\mathbb{P}$ is called the reference probability measure. Define the corresponding natural filtration $(\mathcal{F}_t)_{t=0,1,\ldots,T}$ on Ω through

$$\mathcal{F}_0 := \{\emptyset, \Omega\}, \quad \mathcal{F}_t := \sigma(Y_1, \ldots, Y_t), \quad t = 1, \ldots, T.$$

Furthermore, to model the incomplete information, we assume that all generating random variables Y_t, $t = 1, \ldots, T$ depend on a parameter $\vartheta \in \Theta \subset \mathbb{R}$ which might be unknown and therefore is modelled as a random variable on the given probability space with unknown distribution $\mathcal{L}(\vartheta)$. If ϑ is known, for example with value $\theta \in \Theta$, its distribution reduces to $\mathcal{L}(\vartheta) = \delta_\theta$, where δ_θ denotes the distribution concentrated in θ.

Additionally, we make the following structural assumption.

Assumption 1. *Under ϑ, the random variables $Y_1, \ldots, Y_T$ are independent. Recall that this means that it holds for all $k \in \{1, \ldots, T\}$ and $1 \le t_1 < \cdots < t_k \le T$:*

$$\mathcal{L}(Y_{t_1}, \ldots, Y_{t_k} \mid \vartheta = \theta) = \bigotimes_{l=1}^{k} \mathcal{L}(Y_{t_l} \mid \vartheta = \theta), \quad \theta \in \Theta.$$

Let $\mathcal{P}(\Theta)$ be the set of all probability measures on Θ so that we have $\mathcal{L}(\vartheta) \in \mathcal{P}(\Theta)$, and equip $\mathcal{P}(\Theta)$ with the standard σ-algebra $\mathcal{M}_\Theta$ generated by the so-called evaluation-functions $\tau_B : \mathcal{P}(\Theta) \to [0, 1]$ defined for every $B \in \mathcal{B}_\Theta$ through

$$\tau_B(\mu) := \mu(B), \quad \mu \in \mathcal{P}(\Theta).$$

Formally, we have $\mathcal{M}_\Theta := \sigma(\{\tau_B^{-1}(A) \mid A \in \mathcal{B}_{[0,1]}, B \in \mathcal{B}_\Theta\})$.

Our aim is to introduce dynamic risk measures for $(\mathcal{F}_t)_{t\in\mathcal{T}}$-adapted income processes $I = (I_t)_{t=1,\dots,T}$ which incorporate this structure with an unknown parameter. The set of all $(\mathcal{F}_t)_{t=0,1,\dots,T}$–adapted and integrable processes is again denoted by $\mathcal{X}$. More generally than in Chapter 3, we call any family $(\rho_t)_{t=0,1,\dots,T-1}$ of mappings $\rho_t : \mathcal{X} \times \Omega \to \mathbb{R}$ a dynamic risk measure, if, for fixed $I \in \mathcal{X}$, $(\rho_t(I))_{t=0,1,\dots,T-1}$ is an $(\mathcal{F}_t)_{t\in\mathcal{T}}$-adapted stochastic process.

Here, we only consider a special class of income processes. To specify our model, assume that we are given a process $(Z_t)_{t=0,1,\dots,T}$ through

$$Z_0 \in \mathbb{R}, \quad Z_t := g_t(Z_{t-1}, Y_t), \quad t = 1, \dots, T,$$

where $g_t : \mathbb{R}^2 \to \mathbb{R}$ are $(\mathcal{B}^2, \mathcal{B})$–measurable functions. Given ϑ, (Z_t) is a (not necessarily homogeneous) Markov chain. Only income processes $I \in \mathcal{X}^{\mathrm{M}}$ are taken into account. Recall that this means that for $t = 1, \dots, T$, the random variable I_t depends only on Z_{t-1} and Y_t , or in other words, there exist $(\mathcal{B}^2, \mathcal{B})$–measurable functions $h_t^I : \mathbb{R}^2 \to \mathbb{R}$, such that

$$I_t = h_t^I(Z_{t-1}, Y_t), \quad t = 1, \dots, T. \tag{5.1}$$

This construction has already sufficiently been motivated in Section 4.3.

5.2 A Bayesian control approach to dynamic risk measures

In this section we generate a class of risk measures by introducing a Markov decision process which is defined for any given income process $I \in \mathcal{X}^{\mathrm{M}}$ and where its solution corresponds to a Bayesian approach, see below. This approach can be helpful when facing model uncertainty contained in the parameter ϑ which was introduced above. In practical problems, the parameter and even its distribution $\mathcal{L}(\vartheta)$ might not be known. To deal with this situation in a dynamic setting, one will chose at time $t = 0$ a certain prior distribution as an estimation of $\mathcal{L}(\vartheta)$ and try to use the gain of information over time to improve this estimation.

The economic interpretation is inspired by the works Pflug and Ruszczyński (2001), Pflug and Ruszczyński (2005) and Pflug (2006). We assume that a company or an investor gets an income after each period $(t-1, t]$ for any $t = 1, \dots, T$ and is also able to consume an amount a_{t-1}, which has to be decided upon at time $t-1$. Hence, we assume a_{t-1} to be $\mathcal{F}_{t-1}$-measurable. Furthermore, we have to incorporate the unknown distribution of the parameter ϑ by extending the state space introduced in Section 4.3.

For every fixed $\theta \in \Theta$, we can formulate an optimization problem as in Chapter 4 by replacing the reference probability $\mathbb{P}$ with $\mathbb{P}(\,\cdot\,|\vartheta = \theta)$, since Assumption 1 ensures that the independence property is fulfilled. In this case, reward functions $(r_t^\theta)_{\theta\in\Theta}$, $t = 0, 1, \dots, T-1$ and $(V_T^\theta)_{\theta\in\Theta}$ have to be considered and we obtain a family of value functions $(V_t^\theta)_{\theta\in\Theta}$ for every $t \in \{0, 1, \dots, T\}$. Then we are able to solve the optimization problem via the value iteration for every $\theta \in \Theta$. Hence, the resulting optimal policy π^* depends in general on the chosen parameter θ. Denote all other functions and spaces occurring in Section 4.3 with a prime.

Instead of using the approach just described we will try to find optimal Bayes-policies for certain models, which do not depend on the underlying parameter θ anymore. This means that we will choose a certain prior distribution $\mu_0 \in \mathcal{P}(\Theta)$ and then search for the solution of

$$V_0^{\mu_0}(s) := \sup_{\pi\in(F^T)'} \int_\Theta V_{0,\pi}^\theta(s)\mu_0(\mathrm{d}\theta), \quad s \in S'. \tag{5.2}$$

We now formulate a general non-stationary Bayesian control model which solves this optimization problem and then show how to derive a dynamic risk measure for income processes by this approach.

In the first step, let us introduce all necessary quantities well-known from Markov decision theory. After that, we define the corresponding risk measure, which derives easily from the resulting value functions. The notation is analogously to that in Rieder (1975), for example.

The income and consumption model described above generates a wealth process $(W_t)_{t=0,1,\dots,T}$, where its values depend on W_{t-1}, I_t and a decision variable a_{t-1} for every $t = 1, \dots, T$. More precisely, assume $(W_t)_{t=0,1,\dots,T}$ that is of the form

$$W_0 \equiv 0, \quad W_t = F_t(W_{t-1}, I_t, a_{t-1}), \quad t = 1, \dots, T.$$

with given measurable transition functions $F_t : \mathbb{R}^3 \to \mathbb{R}$.

We are now able to start with the formulation of our Markov decision process. Let $I \in \mathcal{X}^{\mathrm{M}}$ be fixed.

- The *state space* is denoted by $S \subset \mathbb{R}^2 \times \mathcal{P}(\Theta)$ and equipped with the corresponding σ algebra $\mathcal{S} := (\mathcal{B}^2 \otimes \mathcal{M}_\Theta)_S$. For convenience, we sometimes use the notation
$$S = S' \times M = S_1 \times S_2 \times M.$$
Let $s := (s', \mu) := (w, z, \mu) \in S$ be an element of the state space, where w, z, μ represent realizations of the wealth process W_t, the generating Markov chain Z_t and the current estimation of the distribution of ϑ at time t, respectively.

- The *action space* is $A \subset \mathbb{R}$ equipped with $\mathcal{A} := \mathcal{B}_A$. Then $a \in A$ denotes the invested amount.

- In our models, there are no restrictions on the actions, therefore for state $s \in S$ the space of *admissible policies* is $D(s) = A$ and the *restriction set* is $D = S \times A \subset \mathbb{R}^2 \times \mathcal{P}(\Theta) \times \mathbb{R}$ equipped with the σ Algebra $D = \mathcal{S} \otimes \mathcal{A}$.

- The *noise* has values in $E \subset \mathbb{R}$ equipped with $\mathcal{E} := \mathcal{B}_E$.
- The *transition function* $T_t : D \times E \to S$ at time $t = 1, \dots, T$ is given by

  $$T_t(s, a, y) := (T'_t(s, a, y), \Phi_t(s, a, T'_t(s, a, y)))$$

  with

  $$T'_t(s, a, y) = (F_t(w, h^I_t(z, y), a), g_t(z, y))$$

  for $(s, a, y) = (w, z, \mu, a, y) \in D \times E$, where $\Phi_t : D \times S' \to \mathcal{P}(\Theta)$ is the so-called *Bayes operator*, which updates the estimated distribution $\mu \in \mathcal{P}(\Theta)$ by using the noise $y \in E$. It will be defined and further investigated below.
- The *transition law* $Q_t : D \times \mathcal{S} \to [0, 1]$ at time $t = 1, \dots, T$ is the probability kernel between $(D, \mathcal{D})$ and $(S, \mathcal{S})$ defined by

  $$Q_t(s, a; B) := \int\int \mathbb{1}_B(x, \Phi_t(s, a, x)) Q^\theta_t(s, a; \mathrm{d}x)\mu(\mathrm{d}\theta), \quad B \in \mathcal{S},$$

  for $(s, a) \in D$, where $Q^\theta_t(s, a; A) = \mathbb{P}(T'_t(s, a, Y_t) \in A \,|\, \vartheta = \theta)$, $A \in \mathcal{S}'$, and, by definition, $\mathbb{P}(T'_t(s, a, Y_t) \in A \,|\, \vartheta = \cdot)$ is a $(\mathcal{B}_\Theta, \mathcal{B}_{[0,1]})$-measurable mapping for every $A \in \mathcal{S}'$.
- The *one-step reward function* at time $t = 0, 1, \dots, T - 1$ is the measurable mapping $r_t : D \to \mathbb{R}$ defined via

  $$r_t(s, \mu, a) = \int_\Theta r^\theta_t(s, a)\, \mu(\mathrm{d}\theta), \quad (s, \mu, a) \in D.$$
- The *terminal reward function* is the measurable mapping $V_T : S \to \mathbb{R}$ defined via

  $$V_T(s, \mu) = \int_\Theta V^\theta_T(s)\, \mu(\mathrm{d}\theta), \quad (s, \mu) \in S.$$

Remark. Inserting the random quantities W_{t-1}, Z_{t-1}, a_{t-1}, Y_t and $\mu_{t-1} \in \mathcal{P}(\Theta)$ and furthermore defining $\mu_t := \Phi_t(W_{t-1}, Z_{t-1}, \mu_{t-1}, a_{t-1}, W_t, Z_t)$, we obtain for the value of the transition function

$$\begin{aligned} &T_t(W_{t-1}, Z_{t-1}, \mu_{t-1}, a_{t-1}, Y_t) \\ &= (F_t(W_{t-1}, h^I_t(Z_{t-1}, Y_t), a_{t-1}), g_t(Z_{t-1}, Y_t), \mu_t) \\ &= (W_t, Z_t, \mu_t). \end{aligned}$$

By this, a classic Markov decision process as investigated in Hernández-Lerma and Lasserre (1996) with $X_t = (W_t, Z_t, \mu_t)$ is defined.

A chosen initial distribution $\mu_0 \in \mathcal{P}(\Theta)$ is called the *prior distribution*, while the μ_t, $t = 1, \dots, T$, are called *posterior distributions*. They can be interpreted as the distribution of ϑ given the history of the Markov decision process at time t if the true distribution of ϑ is μ_0.

As mentioned above, let us further describe the Bayes operator. Similar to Rieder (1975), we make the following assumption.

Assumption 2. *There exists a σ-finite measure ν on $\mathcal{S}'$, such that for every $t \in \{1, \ldots, T\}$ and $(s, a) \in D'$ the transition law*

$$Q_t^\theta(s, a; \cdot) = \mathbb{P}(T_t'(s, a, Y_t) \in \cdot \mid \vartheta = \theta)$$

has a density q_t with respect to ν, that is we assume

$$\mathbb{P}(T_t'(s, a, Y_t) \in \mathrm{d}x \mid \vartheta = \theta) = q_t^\theta(x \mid s, a)\nu(\mathrm{d}x), \quad \theta \in \Theta.$$

The Bayes operator $\Phi_t : D \times \mathcal{S}' \to \mathcal{P}(\Theta)$ is now defined for $t = 1, \ldots, T$ via

$$\Phi_t(s, \mu, a, x)(B) := \frac{\int_B q_t^\theta(x \mid s, a)\mu(\mathrm{d}\theta)}{\int_\Theta q_t^\theta(x \mid s, a)\mu(\mathrm{d}\theta)}, \quad B \in \mathcal{B}_\Theta, \ (s, \mu, a, x) \in D \times S'.$$

Example 5.1. Let us see how the transition law and the Bayes operator look like if we assume that the generating random variables follow a two-point distribution, i. e. set $E := \{u, d\}$, $\Theta := [0, 1]$ and

$$\mathbb{P}(Y_t = u \mid \vartheta = \theta) = \theta = 1 - \mathbb{P}(Y_t = d \mid \vartheta = \theta), \quad \theta \in \Theta, \ t = 1, \ldots, T.$$

Furthermore, let $(s, \mu, a) \in S$ be the state at time $t - 1$ with $\mu \notin \{\delta_\theta \mid \theta \in [0, 1]\}$, such that in particular $0 < m_\mu := \int_\Theta \theta\mu(\mathrm{d}\theta) < 1$. Assume $T_t'(s, a, u) \neq T_t'(s, a, d)$. Then, ν (compare Assumption 2) becomes the counting measure on $\mathcal{S}'$. Hence, there are two possible states (for the part in S') at time t, namely $x_u := T_t'(s, a, u)$ and $x_d := T_t'(s, a, d)$. We obtain

$$q_t^\theta(x_y \mid s, a) = \mathbb{P}(T_t'(s, a, Y_t) = x_y \mid \vartheta = \theta) = \mathbb{P}(Y_t = y \mid \vartheta = \theta), \quad y \in E, \theta \in \Theta.$$

Consequently, the Bayes operator only has to be calculated for $x \in \{x_u, x_d\} \subset S'$ and becomes for $B \in \mathcal{B}_{[0,1]}$

$$\mu_u(B) := \Phi_t(s, \mu, a, x_u)(B) = \frac{\int_B q_t^\theta(x_u \mid s, a)\mu(\mathrm{d}\theta)}{\int_\Theta q_t^\theta(x_u \mid s, a)\mu(\mathrm{d}\theta)} = \frac{\int_B \theta\,\mu(\mathrm{d}\theta)}{\int_\Theta \theta\,\mu(\mathrm{d}\theta)} = \frac{1}{m_\mu}\int_B \theta\,\mu(\mathrm{d}\theta)$$

and analogously

$$\mu_d(B) := \Phi_t(s, \mu, a, x_d)(B) = \frac{1}{1 - m_\mu}\int_B (1 - \theta)\,\mu(\mathrm{d}\theta).$$

We conclude that the transition law also follows a two-point distribution, namely on

$$G := \{(x_u, \mu_u), (x_d, \mu_d)\}.$$

It is now easily calculated via

$$\begin{aligned} Q_t(s,\mu,a;\{x_u\}\times\{\mu_u\}) &= \int\int \mathbb{1}_{\{x_u\}\times\{\mu_u\}}(x,\mu_u)Q_t^\theta(s,a;\mathrm{d}x)\mu(\mathrm{d}\theta) \\ &= \int Q_t^\theta(s,a;\{x_u\})\mu(\mathrm{d}\theta) = \int q_t^\theta(x_u\,|\,s,a)\mu(\mathrm{d}\theta) \\ &= \int \theta\mu(\mathrm{d}\theta) = m_\mu \end{aligned}$$

and similarly $Q_t(s,\mu,a;\{x_d\}\times\{\mu_d\}) = 1 - m_\mu$. Therefore,

$$Q_t(s,\mu,a;\{x_y\}\times\{\mu_y\}) = \mathbb{P}(Y_t = y\,|\,\vartheta = m_\mu), \quad y\in E, \tag{5.3}$$

concluding the example.

With the variables defined, we have to maximize the expected reward over the periods 1 to T, that is the term

$$\mathbb{E}\left[\sum_{k=0}^{T-1} r_k(X_k,a_k) + V_T(X_T)\,\middle|\, X_0 = (s,\mu)\right]$$

for fixed $(s,\mu)\in S$ over a set of given sequences $(a_0,\dots,a_{T-1})$ of decisions. These are specified by the following definition.

Definition 5.1. *For $t = 0,\dots,T-1$, the set of $(T-t)$–step* admissible Markov policies *is*

$$F^{T-t} := \{\pi = (f_t,\dots,f_{T-1})\,|\, f_k : S\to A\ (\mathcal{S},\mathcal{A})\text{-meas.}, k = t,\dots,T-1\}.$$

Now, we are ready to introduce the sequence of *value functions* $(V_t)_{t=0,1,\dots,T}$.

Fix $t\in\{0,1,\dots,T-1\}$ and $(s,\mu)\in S$. For given $\pi\in F^{T-t}$ the expected reward over the periods $t+1$ to T is

$$V_{t,\pi}(s,\mu) := \mathbb{E}\left[\sum_{k=t}^{T-1} r_k(X_k,f_k(X_k)) + V_T(X_T)\,\middle|\, X_t = (s,\mu)\right]$$

and the value function V_t is the maximal expected reward over all $\pi\in F^{T-t}$, consequently

$$V_t(s,\mu) := \sup_{\pi\in F^{T-t}} V_{t,\pi}(s,\mu).$$

Now, the task is to solve this optimization problem and to establish if there is an optimal policy, such that the supremum is a maximum, that is if there is a $\pi^*\in F^{T-t}$ such that

$$V_t(s,\mu) := V_{t,\pi^*}(s,\mu).$$

Analogously to the case $t = 0$, we set for $t \in \{0, 1, \dots, T-1\}$ and $\mu \in \mathcal{P}(\Theta)$

$$V_t^{\mu}(s) := \sup_{\pi \in (F^{T-t})'} \int_{\Theta} V_{t,\pi}^{\theta}(s)\mu(\mathrm{d}\theta), \quad s \in S'.$$

Then, the following result (see Theorem 7.3 in Rieder (1975)) is well known. For a similar thorough treatment of the subject, compare also Chapter 3 in van Hee (1978).

Theorem 5.1. *Assume that either*

$$\sup_{\pi \in F^{T-t}} \mathbb{E}\left[\sum_{k=t}^{T-1} r_k^{+}(X_k, f_k(X_k)) + V_T^{+}(X_T) \,\Big|\, X_t = (s,\mu)\right] < \infty, \quad (s,\mu) \in S,$$

or

$$\sup_{\pi \in F^{T-t}} \mathbb{E}\left[\sum_{k=t}^{T-1} r_k^{-}(X_k, f_k(X_k)) + V_T^{-}(X_T) \,\Big|\, X_t = (s,\mu)\right] < \infty, \quad (s,\mu) \in S.$$

Then for all $t \in \{0, 1, \dots, T\}$, it holds

$$V_t(s,\mu) = V_t^{\mu}(s), \quad (s,\mu) \in S,$$

In this way, we can solve (5.2) by considering an ordinary Markov decision problem with an extended state space (compared to Section 4.3. To derive the optimal policy and explicit representations of the value functions we can use again Theorem 3.2.1 from Hernández-Lerma and Lasserre (1996). For the sake of completeness, we give here the reformulation of Theorem 4.1 in the context of our model.

Theorem 5.2. *Let $t \in \{1, \dots, T\}$ and $(s,\mu) \in S$ be fixed. If the functions $J_T(s',\mu') := V_T(s',\mu')$ and*

$$J_k(s',\mu') := \sup_{a \in A} \left\{ r_k(s',\mu',a) + \mathbb{E}\left[J_{k+1}(X_{k+1}) \,\big|\, X_k = (s',\mu'),\, a_k = a \right] \right\} \tag{5.4}$$

defined for $k = t, \dots, T-1$ are measurable and if moreover the supremum is attained in $a^* = f_k(s',\mu')$, such that $f_k : S \to A$ is an $(\mathcal{S}, \mathcal{A})$ measurable function, then $\pi^* := (f_t, \dots, f_{T-1})$ is an optimal policy in the sense that

$$V_t(s,\mu) = V_{t,\pi^*}(s,\mu) = J_t(s,\mu).$$

Remark. Again, we can rewrite the *value iteration* (5.4) in terms of the value functions for $t = 0, \dots, T-1$:

$$V_t(s,\mu) := \max_{a \in A} \left\{ r_t(s,\mu,a) + \mathbb{E}\left[V_{t+1}(X_{t+1}) \,\big|\, X_t = (s,\mu),\, a_t = a \right] \right\}. \tag{5.5}$$

Now, we are able to introduce our dynamic risk measure, which is motivated by (4.2).

Definition 5.2. *Consider a Markov decision model for an Markov income process $I = (I_t)_{t=1,\dots,T} \in \mathcal{X}^M$ as described in Section 5.2 and let the assumptions of Theorem 5.2 be fulfilled. Furthermore, assume that we are given a wealth process $(W_t)_{t=0,1,\dots,T}$ and an initial distribution $\mu_0 \in \mathcal{P}(\Theta)$ from which we obtain the sequence of posterior distributions $(\mu_t)_{t=1,\dots,T}$, and define*

$$X_t := (W_t, Z_t, \mu_t), \quad t = 0, 1, \dots, T.$$

Introduce measurable correction functions $v_t : S_1 \to \mathbb{R}$ for $t \in \{0, 1, \dots, T-1\}$ and define a dynamic risk measure via

$$\rho_t^{\mathrm{B},\mu_0}(I) := v_t(W_t) - \frac{1}{c_t} V_t(X_t), \quad t = 0, 1, \dots, T-1.$$

Remark.
1. Since this definition requires the assumptions of Theorem 5.2 to be fulfilled, every $V_t(\cdot)$ is a measurable function and therefore every $\rho_t^{\mathrm{B}}(I)$ is $\mathcal{F}_t$-measurable. Consequently, the dynamic risk measure $(\rho_t^B(I))_{t=0,1,\dots,T-1}$ is indeed an $(\mathcal{F}_t)_{t=0,1,\dots,T-1}$-adapted process.

2. As initial wealth process we usually choose the optimal one, i. e. we solve the optimization problem at time $t = 0$ and obtain an optimal Markov policy $\pi^* = (f_0^*, \dots, f_{T-1}^*) \in F^T$. With this in hand, we set for some $W_0 \in S_1$
$$W_t = F_t(W_{t-1}, I_t, f_{t-1}^*(W_{t-1}, Z_{t-1}, \mu_{t-1})), \quad t = 1, \dots, T.$$

3. Looking at (4.2) motivates the introduction of the correction functions. In that example, which we will use throughout the following section, we have $v_t(w) = -\frac{q_t}{c_t} w^-$, $w \in S_1$.

We have introduced a new class of dynamic risk measures and summarized the necessary results from the literature. For every choice of the reward functions, a different risk measure is generated. In the following section, we will derive explicit formulas by using the functions which occurred in Chapter 4.

5.3 Explicit solutions

In this section, we want to solve the optimization problem introduced in the previous section for certain initial distributions and for the reward functions used in Chapter 4, namely:

- The transition function becomes for $t = 1, \dots, T$
$$T_t(s, \mu, a, y) := (T_t'(s, a, y), \Phi_t(\mu, y)), \quad (s, \mu, a, y) = (w, z, \mu, a) \in D \times E,$$
where $T_t'(s, a, y) = (w^+ + h_t(z, y) - a, g_t(z, y))$.

- The reward function becomes for $t = 0, \dots, T-1$

$$r'_t(s,a) := r_t(s,\mu,a) = -q_t w^- + c_{t+1}a, \quad (s,\mu,a) = (w,z,\mu,a) \in D.$$

where we set for convenience $q_0 := 0$.

- The terminal reward function becomes

$$V'_T(s) := V_T(s,\mu) = c_{T+1}w^+ - q_T w^-, \quad (s,\mu) = (w,z,\mu) \in S.$$

As initial distributions, we use certain classes of distributions which are preserved under the Bayes operator, e. g. one point distributions, which usage corresponds to the case of a known parameter ϑ, and, as an example of continuous distributions, the Beta distributions. In the first case, we can treat the model introduced in this chapter as in Section 4.3. The other example interprets the parameter ϑ as an unknown probability.

5.3.1 The case when the parameter ϑ is known

Here, we assume that $\mathbb{P}(\vartheta = \theta_0) = 1$ for some known $\theta_0 \in \Theta$, i. e. set $\mu_0 = \delta_{\theta_0}$. In the first step, we calculate the resulting Bayes operator and the transition kernel. This gives for all $t \in \{1, \dots, T\}$

$$\Phi_t(s,\delta_{\theta_0},a,x)(B) = \frac{\int_B q_t^\theta(x \mid s,a)\delta_{\theta_0}(\mathrm{d}\theta)}{\int_\Theta q_t^\theta(x \mid s,a)\delta_{\theta_0}(\mathrm{d}\theta)} = \frac{q_t^{\theta_0}(x \mid s,a)\mathbb{1}_B(\theta_0)}{q_t^{\theta_0}(x \mid s,a)} = \mathbb{1}_B(\theta_0), \ B \in \mathcal{B}_\Theta,$$

for all $s \in S'$, $a \in A$ and $x \in S'$, so that we obtain inductively $\mu_t = \delta_{\theta_0}$ for all $t \in \{1, \dots, T\}$, independent of the history of the Markov decision process. We conclude that the estimated distribution of ϑ, starting with δ_{θ_0}, does not change over time. Furthermore,

$$\begin{aligned} Q_t(s,\delta_{\theta_0},a;B' \times \{\delta_{\theta_0}\}) &= \int\int \mathbb{1}_{B' \times \{\delta_{\theta_0}\}}(x,\delta_{\theta_0}) Q_t^\theta(s,a;\mathrm{d}x)\delta_{\theta_0}(\mathrm{d}\theta) \\ &= Q_t^{\theta_0}(s,a;B'), \quad (s,a) \in D', \quad B' \in \mathcal{S}', \end{aligned}$$

which is just the transition law used in Section 4.3 when the reference probability $\mathbb{P}$ is replaced by $\mathbb{P}(\,\cdot \mid \vartheta = \theta_0)$. Furthermore, $Q_t(s,\delta_{\theta_0},a;\cdot)$ is concentrated on $S' \times \{\delta_{\theta_0}\}$ such that we can indeed identify $Q_t(s,\delta_{\theta_0},a;\,\cdot \times \{\delta_{\theta_0}\})$ with $Q_t^{\theta_0}(s,a;\,\cdot\,)$. Summing things up, the dynamic risk measure ρ^{B} is the same as the dynamic risk measure ρ^{PR} with the new reference probability. So, if we denote with $\mathbb{E}^{\theta_0}$ the expectation, with $\mathrm{V@R}^{\theta_0}$ the Value-at-Risk and with $\mathrm{AV@R}^{\theta_0}$ the Average Value-at-Risk with respect to $\mathbb{P}^{\theta_0} := \mathbb{P}(\,\cdot \mid \vartheta = \theta_0)$, we can define for $k = 1, \dots, T$

$$\rho_{\theta_0}^{(k)}(X) := \lambda_k \mathbb{E}^{\theta_0}[-X] + (1-\lambda_k)\mathrm{AV@R}_{\gamma_k}^{\theta_0}(X), \quad X \in L^1(\Omega,\mathcal{F},\mathbb{P}),$$

and obtain for $t \in \{0,1,\dots,T-1\}$ and $I \in \mathcal{X}^{\mathrm{M}}$

$$\begin{aligned}\rho_t^{\mathrm{B},\delta_{\theta_0}}(I) &= \rho_t^{\mathrm{PR},\theta_0}(I)\\ &= -\frac{c_{t+1}}{c_t}(I_t + \mathrm{V@R}_{\gamma_t}^{\theta_0}(I_t \,|\, Z_{t-1}))^+ + \mathbb{E}^{\theta_0}\left[\sum_{k=t+1}^{T} \frac{c_k}{c_t}\rho_{\theta_0}^{(k)}(I_k \,|\, Z_{k-1}) \,\Big|\, Z_t\right].\end{aligned}$$

5.3.2 Beta distributions as initial distribution

The result of this subsection can be used to treat e. g. the Artzner–game and the Cox–Ross-Rubinstein–model. These were introduced in Examples 4.1 and 4.2. To derive explicit solutions we implement general Beta distribution as initial distribution and furthermore assume that $\Theta = [0,1]$ and

$$\mathbb{P}(Y_t = u \,|\, \vartheta = \theta) = \theta = 1 - \mathbb{P}(Y_t = d \,|\, \vartheta = \theta), \quad \theta \in \Theta,\ t = 1,\dots,T,$$

for some $0 \le d < 1 \le u$. In this setting, the Beta distribution is a conjugate prior distribution for the binomial distribution (compare e. g. Runggaldier et al. (2002)). If we have no information about the parameter ϑ at time $t = 0$, we usually start with $\mu_0 = \mathcal{U}(0,1)$, which is a special Beta distribution, namely $\mathcal{U}(0,1) = \mathrm{Beta}(1,1)$ in our notation.

To solve the value iteration we again have to calculate the Bayes operator and the transition kernel. It is well known that the Bayes operator preserves the class of Beta distributions, which can easily be seen as follows. Let $t \in \{1,\dots,T\}$ and $\mu = \mathrm{Beta}(\alpha,\beta)$, $\alpha,\beta \ge 0$. By Example 5.1 and (B.1), we have for $B \in \mathcal{B}_{[0,1]}$ and $(s,a) \in D'$, $x_u := T_t'(s,a,u)$

$$\begin{aligned}\Phi_t(s,\mu,a,x_u)(B) &= \frac{\int_B \theta\mu(\mathrm{d}\theta)}{\int_\Theta \theta\mu(\mathrm{d}\theta)} = \frac{\int_B \theta\,\theta^{\alpha-1}(1-\theta)^{\beta-1}(B(\alpha,\beta))^{-1}\mathrm{d}\theta}{m_\mu}\\ &= \frac{\int_B \theta^{\alpha}(1-\theta)^{\beta-1}\mathrm{d}\theta}{B(\alpha+1,\beta)} = \mathrm{Beta}(\alpha+1,\beta)(B),\end{aligned}$$

i. e. $\Phi_t(s,\mu,a,x_u) = \mathrm{Beta}(\alpha+1,\beta)$ and analogously

$$\Phi_t(s,\mu,a,x_d) = \mathrm{Beta}(\alpha,\beta+1)$$

with $x_d := T_t'(s,a,d)$. Combining these observations, it follows

$$\Phi_t(s,\mathrm{Beta}(\alpha,\beta),a,T_t'(s,a,Y_t)) = \mathrm{Beta}(\alpha + \mathbb{1}_{\{u\}}(Y_t), \beta + \mathbb{1}_{\{d\}}(Y_t)). \tag{5.6}$$

Consequently, with such an initial distribution, we only need to calculate the transition kernel for this class of distributions. Again, Example 5.1 yields

$$Q_t(s,\mu,a;\{x_u\} \times \{\mathrm{Beta}(\alpha+1,\beta)\}) = m_\mu = \frac{\alpha}{\alpha+\beta} \tag{5.7}$$

and analogously

$$Q_t(s,\mu,a;\{x_d\}\times\{\mathrm{Beta}(\alpha,\beta+1)\}) = 1-m_\mu = \frac{\beta}{\alpha+\beta}. \tag{5.8}$$

Hence, $Q_t(s,\mu,a;\cdot)$ is a two point distribution on

$$G := \{(x_u, \mathrm{Beta}(\alpha+1,\beta)), (x_d, \mathrm{Beta}(\alpha,\beta+1))\}$$

Now, we are ready to solve the value iteration. To avoid some notational difficulties, we introduce some abbreviations. Write the state space as $S = S_1\times S_2\times M$ with

$$M := \{\mathrm{Beta}(\alpha,\beta)\,|\,\alpha,\beta\geq 0\}$$

and define for $t\in\{1,\dots,T\}$ and $k\in\{t,\dots,T\}$ the functions

$$\begin{aligned} g_k^t &: S_2\times E^{k-t}\to\mathbb{R} \\ (s_2; y_{t+1},\dots,y_k) &\mapsto g_k(g_{k-1}(\dots(g_{t+1}(s_2,y_{t+1})\dots),y_{k-1}),y_k). \end{aligned}$$

Note that we get the special cases $g_t^t = \mathrm{id}_{S_2}$ and $g_{t+1}^t = g_t$ for $k=t$ and $k=t+1$ respectively and that we obtain

$$\mathcal{L}(g_k^t(s_2;Y_{t+1},\dots,Y_k)) = \mathcal{L}(Z_k\,|\,Z_t = s_2), \quad s_2\in S_2.$$

Furthermore, for $k=t,\dots,T$ we have with $(s_2;y_t,y_{t+1},\dots,y_k)\in S_2\times E^{k-t+1}$

$$g_k^t(g_t(s_2,y_t);y_{t+1},\dots,y_k) = g_k^{t-1}(s_2;y_t,y_{t+1},\dots,y_k). \tag{5.9}$$

Recall that for $k\in\{1,\dots,T\}$ and $\gamma\in[0,1]$ we introduced the static coherent risk measure

$$\rho^{(k)}(X) := \lambda_k\mathbb{E}[-X] + (1-\lambda_k)\mathrm{AV@R}_{\gamma_k}(X), \quad X\in L^1(\Omega,\mathcal{F},\mathbb{P}).$$

Theorem 5.3. *Let $t\in\{0,1,\dots,T\}$, $(w,z)\in S'$. For $\alpha,\beta>0$ we then have the following:*

(i) The value function is given by

$$\begin{aligned} V_t(w,z,\mathrm{Beta}(\alpha,\beta)) &= c_{t+1}w^+ - q_t w^- \\ &- \sum_{k=t+1}^{T} c_k\cdot \sum_{\substack{(z;y)\in\{z\}\times E^{k-(t+1)} \\ y=(y_{t+1},\dots,y_{k-1})}} \prod_{j=t+1}^{k-1} \mathbb{P}\Big(Y_j = y_j\,|\,\vartheta = \frac{\alpha+\sum_{i=t+1}^{j-1}\mathbb{1}_{\{u\}}(y_i)}{\alpha+\beta+j-(t+1)}\Big)\times \\ &\qquad \rho^{(k)}\Big(I_k\,|\,Z_{k-1} = g_{k-1}^t(z;y_{t+1},\dots,y_{k-1}), \vartheta = \frac{\alpha+\sum_{i=t+1}^{k-1}\mathbb{1}_{\{u\}}(y_i)}{\alpha+\beta+k-(t+1)}\Big). \end{aligned}$$

(ii) The optimal policy $\pi^* = (f_t^*, \dots, f_{T-1}^*)$ *is*

$$f_k^*(w', z', \mathrm{Beta}(\alpha, \beta)) = (w')^+ - \mathrm{V@R}_{\gamma_{k+1}}\Big(I_{k+1} \,\big|\, Z_k = z', \vartheta = \frac{\alpha}{\alpha+\beta}\Big)$$

for $k = t, \dots, T-1$ *and* $(w', z') \in S'$. *Moreover, the optimal Markov process* $(X_t^*, \dots, X_T^*)$ *is given via the recursive relation*

$$X_t^* = (w, z)$$

and for $k = t+1, \dots, T$

$$X_k^* = (X_{k,1}^*, X_{k,2}^*, X_{k,3}^*),$$

where

$$\begin{aligned} X_{k,1}^* &= h_k^I(X_{k-1,2}^*, Y_k) \\ &\quad + \mathrm{V@R}_{\gamma_k}\Big(I_k \,\big|\, Z_{k-1} = X_{k-1,2}^*, \vartheta = \frac{\alpha + \sum_{i=t+1}^{k-1} \mathbb{1}_{\{u\}}(Y_i)}{\alpha + \beta + k - t - 1}\Big), \\ X_{k,2}^* &= g_k(X_{k-1,2}^*, Y_k), \\ X_{k,3}^* &= \mathrm{Beta}\Big(\alpha + \sum_{i=t+1}^{k} \mathbb{1}_{\{u\}}(Y_i), \beta + \sum_{i=t+1}^{k} \mathbb{1}_{\{d\}}(Y_i)\Big). \end{aligned}$$

Proof. The proof is by backward induction on $t \in \{0, 1, \dots, T-1\}$. Let $t = T-1$. The value iteration with $\mu := \mathrm{Beta}(\alpha, \beta)$ yields by (5.7) and (5.8)

$$\begin{aligned} &V_{T-1}(w, z, \mu) \\ &= \sup_{a \in A} \big\{ r_{T-1}(w, z, \mu, a) + \mathbb{E}\left[V_T(X_T) \mid X_{T-1} = (w, z, \mu), a_{T-1} = a\right] \big\} \\ &= \sup_{a \in A} \big\{ r_{T-1}(w, z, \mu, a) + \int_G V_T(x) Q_T(w, z, \mu, a; \mathrm{d}x) \big\} \\ &= \sup_{a \in A} \big\{ r'_{T-1}(w, z, a) + \frac{\alpha}{\alpha+\beta} V'_T(T'_T(w, z, a, u)) + \frac{\beta}{\alpha+\beta} V'_T(T'_T(w, z, a, d)) \big\} \\ &\overset{(5.3)}{=} \sup_{a \in A} \Big\{ r'_{T-1}(w, z, a) + \mathbb{E}\Big[V'_T(T'_T(w, z, a, Y_T)) \,\big|\, \vartheta = \frac{\alpha}{\alpha+\beta}\Big] \Big\}. \end{aligned}$$

This optimization problem can be solved as in the proof of Theorem 4.2 and we obtain

$$\begin{aligned} V_{T-1}(w, z, \mu) &= c_T w^+ - q_{T-1} w^- + c_{T+1} \mathbb{E}\Big[h_T(z, Y_T) \,\big|\, \vartheta = \frac{\alpha}{\alpha+\beta}\Big] \\ &\qquad - (c_T - c_{T+1}) \mathrm{AV@R}_{\gamma_T}\Big(h_T(z, Y_T) \,\big|\, \vartheta = \frac{\alpha}{\alpha+\beta}\Big) \\ &= c_T w^+ - q_{T-1} w^- - c_T \cdot \rho^{(T)}\Big(I_T \,\big|\, Z_{T-1} = z, \vartheta = \frac{\alpha}{\alpha+\beta}\Big), \end{aligned}$$

where we used the fact that under $\mathbb{P}^{\cdot \mid \vartheta = \frac{\alpha}{\alpha+\beta}}$, Z_{T-1} and Y_T are independent. So the assertion is true for $t = T-1$ and we note that the supremum is attained in

$$a^* = w^+ - \mathrm{V@R}_{\gamma_T}\Big(I_T \,\Big|\, Z_{T-1} = z, \vartheta = \frac{\alpha}{\alpha+\beta}\Big),$$

from which we see that

$$W_T^* = h_T^I(z, Y_T) + \mathrm{V@R}_{\gamma_T}\Big(I_T \,\Big|\, Z_{T-1} = z, \vartheta = \frac{\alpha}{\alpha+\beta}\Big).$$

Now assume that the assertion holds for $t \in \{1, \dots, T\}$. To avoid lengthy notations introduce for $k = t+1, \dots, T$, $z \in S_2$ and $\alpha, \beta \geq 0$ the term

$$\begin{aligned} A_t(k; z, \alpha, \beta) := & \sum_{\substack{(z;y) \in \{z\} \times E^{k-(t+1)} \\ y = (y_{t+1}, \dots, y_{k-1})}} \prod_{j=t+1}^{k-1} \mathbb{P}\Big(Y_j = y_j \,\Big|\, \vartheta = \frac{\alpha + \sum_{i=t+1}^{j-1} \mathbb{1}_{\{u\}}(y_i)}{\alpha + \beta + j - (t+1)}\Big) \times \\ & \rho^{(k)}\Big(I_s \,\Big|\, Z_{k-1} = g_{k-1}^t(z; y_{t+1}, \dots, y_{k-1}), \vartheta = \frac{\alpha + \sum_{i=t+1}^{k-1} \mathbb{1}_{\{u\}}(y_i)}{\alpha + \beta + k - (t+1)}\Big) \end{aligned}$$

and note that by (5.9)

$$\begin{aligned} & \mathbb{E}\Big[A_t(k; g_t(z, Y_t), \alpha + \mathbb{1}_{\{u\}}(Y_t), \beta + \mathbb{1}_{\{d\}}(Y_t)) \,\Big|\, \vartheta = \frac{\alpha}{\alpha+\beta}\Big] \\ & = \sum_{y_t \in E} \mathbb{P}\Big(Y_t = y_t \,\Big|\, \vartheta = \frac{\alpha}{\alpha+\beta}\Big) A_t(k; g_t(z, y_t), \alpha + \mathbb{1}_{\{u\}}(y_t), \beta + \mathbb{1}_{\{d\}}(y_t)) \\ & = A_{t-1}(k; z, \alpha, \beta). \end{aligned}$$

Together with the value iteration the induction hypothesis yields similar to the case $t = T-1$ for $\mu := \mathrm{Beta}(\alpha, \beta)$

$$\begin{aligned} & V_{t-1}(w, z, \mu) + q_{t-1} w^- \\ & = \sup_{a \in A} \Big\{ q_{t-1} w^- + r'_{t-1}(w, z, a) + \mathbb{E}\Big[V_t(T_t(w, z, a, Y_t)) \,\Big|\, \vartheta = \frac{\alpha}{\alpha+\beta}\Big]\Big\} \\ & = \sup_{a \in A} \Big\{ c_t a + \mathbb{E}\Big[c_{t+1}(F_t(w, z, a, Y_t))^+ - q_t(F_t(w, z, a, Y_t))^- \,\Big|\, \vartheta = \frac{\alpha}{\alpha+\beta}\Big]\Big\} \\ & \quad - \sum_{k=t+1}^{T} c_k \cdot \mathbb{E}\Big[A_t(k; g_t(z, Y_t), \alpha + \mathbb{1}_{\{u\}}(Y_t), \beta + \mathbb{1}_{\{d\}}(Y_t)) \,\Big|\, \vartheta = \frac{\alpha}{\alpha+\beta}\Big]. \end{aligned}$$

Now, the supremum can again be treated as before and we obtain

$$\begin{aligned}
&V_{t-1}(w,z,\mu)\\
&= c_t w^+ - q_{t-1} w^- - c_t \cdot \rho^{(t)}\Big(I_t \,\Big|\, Z_{t-1}=z, \vartheta = \frac{\alpha}{\alpha+\beta}\Big)\\
&\quad - \sum_{k=t+1}^{T} c_k \cdot \mathbb{E}\Big[A_t(k; g_t(z,Y_t), \alpha + \mathbb{1}_{\{u\}}(Y_t), \beta + \mathbb{1}_{\{d\}}(Y_t)) \,\Big|\, \vartheta = \frac{\alpha}{\alpha+\beta}\Big]\\
&= c_t w^+ - q_{t-1} w^- - c_t \cdot A_{t-1}(t; z, \alpha, \beta) - \sum_{k=t+1}^{T} c_k \cdot A_{t-1}(k; z, \alpha, \beta)\\
&= c_t w^+ - q_{t-1} w^- - \sum_{k=t}^{T} c_s \cdot A_{t-1}(k; z, \alpha, \beta),
\end{aligned}$$

which proves part (i). The supremum is attained in

$$a^* = w^+ - \mathrm{V@R}_{\gamma_t}\Big(I_t \,\Big|\, Z_{t-1} = z, \vartheta = \frac{\alpha}{\alpha+\beta}\Big),$$

from which we see that

$$W_t^* = h_t^I(z, Y_t) + \mathrm{V@R}_{\gamma_t}\Big(I_t \,\Big|\, Z_{t-1} = z, \vartheta = \frac{\alpha}{\alpha+\beta}\Big).$$

The formulas for the optimal policy and the wealth process in part (ii) follow again as in Theorem 4.2 by inserting the quantities recursively. The formula for the second component of X^* is trivial and for the last component, namely the estimated distributions of ϑ, we obtain the desired result by looking at (5.6). □

The closed formula for the value functions in Theorem 5.3 is very lengthy. To deal with this drawback, we rewrite the result in a version that is similar to the one of Theorem 4.2. To this extend, we introduce some new probability measures.

Definition 5.3. *Let $t = 0, 1, \dots, T$ and $\alpha, \beta > 0$. Then we define a probability measure $P_{t,\alpha,\beta}$ on $\sigma(Z_t, Y_{t+1}, \dots, Y_T)$ via*

$$\begin{aligned}
&\mathbb{P}_{t,\alpha,\beta}(Z_t = z, Y_{t+1} = y_{t+1}, \dots, Y_T = y_T)\\
&\quad := \mathbb{P}(Z_t = z) \prod_{j=t+1}^{T} \mathbb{P}\Big(Y_j = y_j \,\Big|\, \vartheta = \frac{\alpha + \sum_{i=t+1}^{j-1} \mathbb{1}_{\{u\}}(y_i)}{\alpha + \beta + j - (t+1)}\Big)
\end{aligned}$$

for all $(z; y_{t+1}, \dots, y_T) \in S_2' \times E^{T-t}$.

It is easy to check that this indeed defines a probability measure. Furthermore, we collect some simple but useful properties in a lemma.

Lemma 5.1. *Let $t = 0, 1, \ldots, T$ and $\alpha, \beta > 0$. Then the following holds:*

(i) Z_t is independent of $(Y_{t+1}, \ldots, Y_T)$ under $\mathbb{P}_{t,\alpha,\beta}$ and we have for $z \in S_2'$

$$\mathbb{P}_{t,\alpha,\beta}(Z_t = z) = \mathbb{P}(Z_t = z). \tag{5.10}$$

(ii) Let $k \in \{t+1, \ldots, T\}$. Then we have for all $(y_{t+1}, \ldots, y_k) \in E^{k-t}$

$$\begin{aligned}&\mathbb{P}_{t,\alpha,\beta}(Y_{t+1} = y_{t+1}, \ldots Y_k = y_k)\\ &= \prod_{j=t+1}^{k} \mathbb{P}\Big(Y_j = y_j \,\Big|\, \vartheta = \frac{\alpha + \sum_{i=t+1}^{j-1} \mathbb{1}_{\{u\}}(y_i)}{\alpha + \beta + j - (t+1)}\Big).\end{aligned} \tag{5.11}$$

(iii) Let $T \geq l \geq k \geq t+1$. For all $(z, y_{t+1}, \ldots, y_l) \in S_2' \times E^{l-t}$ we then have

$$\begin{aligned}&\mathbb{P}_{t,\alpha,\beta}(Y_l = y_l, \ldots, Y_k = y_k \,|\, Y_{k-1} = y_{k-1}, \ldots, Y_{t+1} = y_{t+1}, Z_t = z)\\ &= \prod_{j=k}^{l} \mathbb{P}\Big(Y_j = y_j \,\Big|\, \vartheta = \frac{\alpha + \sum_{i=t+1}^{j-1} \mathbb{1}_{\{u\}}(y_i)}{\alpha + \beta + j - (t+1)}\Big).\end{aligned} \tag{5.12}$$

Proof. (i) Obvious from the product form of the definition.

(ii) We only treat the case $k = T-1$ (the case $k = T$ is clear by definition and part (i)), the rest follows analogously by induction. To this extend, let $(y_{t+1}, \ldots, y_{T-1}) \in E^{T-(t+1)}$:

$$\begin{aligned}&\mathbb{P}_{t,\alpha,\beta}(Y_{t+1} = y_{t+1}, \ldots, Y_{T-1} = y_{T-1})\\ &= \sum_{z \in S_2} \sum_{y_T \in E} \mathbb{P}_{t,\alpha,\beta}(Z_t = z, Y_{t+1} = y_{t+1}, \ldots, Y_T = y_T)\\ &= \prod_{j=t+1}^{T-1} \mathbb{P}\Big(Y_j = y_j \,\Big|\, \vartheta = \frac{\alpha + \sum_{i=t+1}^{j-1} \mathbb{1}_{\{u\}}(y_i)}{\alpha + \beta + j - (t+1)}\Big) \times\\ &\qquad\qquad \sum_{y_T \in E} \mathbb{P}\Big(Y_T = y_T \,\Big|\, \vartheta = \frac{\alpha + \sum_{i=t+1}^{T-1} \mathbb{1}_{\{u\}}(y_i)}{\alpha + \beta + T - (t+1)}\Big)\\ &= \prod_{j=t+1}^{T-1} \mathbb{P}\Big(Y_j = y_j \,\Big|\, \vartheta = \frac{\alpha + \sum_{i=t+1}^{j-1} \mathbb{1}_{\{u\}}(y_i)}{\alpha + \beta + j - (t+1)}\Big).\end{aligned}$$

(iii) We just have to insert the definition of $\mathbb{P}_{t,\alpha,\beta}$:

$$\begin{aligned}
&\mathbb{P}_{t,\alpha,\beta}(Y_l = y_l, \dots, Y_k = y_k \mid Y_{k-1} = y_{k-1}, \dots, Y_{t+1} = y_{t+1}, Z_t = z) \\
&= \frac{\mathbb{P}_{t,\alpha,\beta}(Y_l = y_l, \dots, Y_k = y_k, Y_{k-1} = y_{k-1}, \dots, Y_{t+1} = y_{t+1}, Z_t = z)}{\mathbb{P}_{t,\alpha,\beta}(Y_{k-1} = y_{k-1}, \dots, Y_{t+1} = y_{t+1}, Z_t = z)} \\
&= \frac{\prod_{j=t+1}^{l} \mathbb{P}\Big(Y_j = y_j \mid \vartheta = \frac{\alpha+\sum_{i=t+1}^{j-1} \mathbb{1}_{\{u\}}(y_i)}{\alpha+\beta+j-(t+1)}\Big) \cdot \mathbb{P}(Z_t = z)}{\prod_{j=t+t}^{k-1} \mathbb{P}\Big(Y_j = y_j \mid \vartheta = \frac{\alpha+\sum_{i=t+t}^{j-1} \mathbb{1}_{\{u\}}(y_i)}{\alpha+\beta+j-(t+1)}\Big) \cdot \mathbb{P}(Z_t = z)} \\
&= \prod_{j=k}^{l} \mathbb{P}\Big(Y_j = y_j \mid \vartheta = \frac{\alpha + \sum_{i=t+1}^{j-1} \mathbb{1}_{\{u\}}(y_i)}{\alpha + \beta + j - (t+1)}\Big).
\end{aligned}$$

□

This lemma provides a more intuitive version of the formula in Theorem 5.3. Denote for fixed $t = 0, 1, \dots, T$ and $\alpha, \beta > 0$ by $\mathbb{E}_{t,\alpha,\beta}$ and $\rho^{(k)}_{t,\alpha,\beta}$, $k > t$, the expectation and the risk measure $\rho^{(k)}$ with respect to $\mathbb{P}_{t,\alpha,\beta}$.

Proposition 5.1. *Let $t = 0, 1, \dots, T$ and $\alpha, \beta > 0$. Then for $(w, z) \in S'$*

$$\begin{aligned}
&V_t(w, z, \mathrm{Beta}(\alpha, \beta)) \\
&= c_{t+1} w^+ - q_t w^- - \sum_{k=t+1}^{T} c_k \cdot \mathbb{E}_{t,\alpha,\beta}\Big[\rho^{(k)}_{t,\alpha,\beta}\Big(I_k \mid Y_{k-1}, \dots, Y_{t+1}, Z_t\Big) \mid Z_t = z\Big].
\end{aligned}$$

Proof. First observe that

$$\begin{aligned}
&\rho^{(k)}\Big(I_k \mid Z_{k-1} = g^t_{k-1}(z; y_{t+1}, \dots, y_{k-1}), \vartheta = \frac{\alpha + \sum_{i=t+1}^{k-1} \mathbb{1}_{\{u\}}(y_i)}{\alpha + \beta + k - (t+1)}\Big) \\
&= \rho^{(k)}\Big(h^I_k(g^t_{k-1}(z; y_{t+1}, \dots, y_{k-1}), Y_k) \mid \vartheta = \frac{\alpha + \sum_{i=t+1}^{k-1} \mathbb{1}_{\{u\}}(y_i)}{\alpha + \beta + k - (t+1)}\Big) \\
&\overset{(5.12)}{=} \rho^{(k)}_{t,\alpha,\beta}\Big(h^I_k(g^t_{k-1}(z; y_{t+1}, \dots, y_{k-1}), Y_k) \mid Y_{k-1} = y_{k-1}, \dots, Y_{t+1} = y_{t+1}, Z_t = z\Big) \\
&= \rho^{(k)}_{t,\alpha,\beta}\Big(I_k \mid Y_{k-1} = y_{k-1}, \dots, Y_{t+1} = y_{t+1}, Z_t = k\Big)
\end{aligned}$$

where in the first step we used the fact that $I_s = h^I_s(Z_{s-1}, Y_s)$ by definition and that Y_s and Z_{s-1} are independent under ϑ. Together with Lemma 5.1 this yields

$$\begin{aligned}
&V_t(w, z, \mathrm{Beta}(\alpha, \beta)) \\
&\overset{(5.11)}{=} c_{t+1} w^+ - q_t w^- - \sum_{k=t+1}^{T} c_k \cdot \mathbb{E}_{t,\alpha,\beta}\Big[\rho^{(k)}_{t,\alpha,\beta}\Big(I_k \mid Y_{k-1}, \dots, Y_{t+1}, Z_t = z\Big)\Big] \\
&\overset{(5.10)}{=} c_{t+1} w^+ - q_t w^- - \sum_{k=t+1}^{T} c_k \cdot \mathbb{E}_{t,\alpha,\beta}\Big[\rho^{(k)}_{t,\alpha,\beta}\Big(I_k \mid Y_{k-1}, \dots, Y_{t+1}, Z_t\Big) \mid Z_t = z\Big],
\end{aligned}$$

hence the assertion. □

The representation of the value functions in Proposition 5.1 is now of the same structure as the one in the case where ϑ is known, compare Theorem 4.2. This shall conclude this section. In the next one we will give a result how these two value functions can be ordered. In fact, under some additional assumption, the value function starting with a general distribution from $\mathcal{P}(0,1)$ is always not smaller than the one starting with a one-point distribution for all $(w,z) \in S'$. After looking at the theoretical results, we will give an economic interpretation of this observation.

5.4 Comparison of value functions

In this section we again consider the binomial model, i. e. we assume $\Theta = [0,1]$ and

$$\mathbb{P}(Y_t = u \mid \vartheta = \theta) = \theta = 1 - \mathbb{P}(Y_t = d \mid \vartheta = \theta), \quad \theta \in [0,1],\ t = 1,\dots,T.$$

We have already seen that

$$q_t^\theta(T_t'(s,a,y) \mid s,a) = \mathbb{P}(Y_t = y \mid \vartheta = \theta), \quad y \in \{u,d\}, \theta \in \Theta.$$

Hence, the Bayes operator does not depend on $(s,a) \in D'$. Therefore, we shortly write

$$\Phi_t(\mu, y) := \Phi_t(s,\mu,a,T_t'(s,a,y)), \quad (s,\mu,a) \in D, y \in \{u,d\}.$$

After a few preparations we are ready to prove the main theorem of this section, which we will apply to our main models, if possible.

5.4.1 A comparison result for general distributions

As described at the end of the last section, our main aim is to derive a comparison result for the two risk measures ρ^{PR} and ρ^{B}, where the most important case is the point of time $t = 0$. Formally, we prove that for any initial distribution $\mu \in \mathcal{P}(0,1)$

$$\rho_0^{\text{PR},m_\mu}(I) \geq \rho_0^{\text{B},\mu}(I)$$

for all processes $I \in \mathcal{X}^{\text{M}}$ that are constructed as described in Section 5.1. For any distribution μ on $\mathbb{R}$, we denote, if they exist, by $m_\mu^{(k)}$, $k \in \mathbb{N}$ and σ_μ^2 its k-th moment and its variance respectively. For convenience, set $m_\mu := m_\mu^{(1)}$.

Let us first make a simple observation.

Proposition 5.2. *Let $t \in \{0,1,\dots,T-1\}$ and $s \in S'$. If $\theta \mapsto V_{t,\pi}^\theta(s)$ is convex on $[0,1]$ for every $\pi \in (F^{T-t})'$, then*

$$V_t(s,\mu) \geq V_t(s,\delta_{m_\mu}), \quad \mu \in \mathcal{P}(0,1).$$

Proof. Let $\mu \in \mathcal{P}(0,1)$. Since $\theta \to V^{\theta}_{t,\pi}(s)$ is convex, Jensen's inequality yields

$$\int_{[0,1]} V^{\theta}_{t,\pi}(s)\,\mu(\mathrm{d}\theta) \geq V^{m_\mu}_{t,\pi}(s) = V_{t,\pi}(s,\delta_{m_\mu}), \quad \pi \in (F^{T-t})'. \tag{5.13}$$

Consequently,

$$V_t(s,\mu) = \sup_{\pi\in(F^{T-t})'} \int_{[0,1]} V^{\theta}_{t,\pi}(s)\,\mu(\mathrm{d}\theta) \overset{(5.13)}{\geq} \sup_{\pi\in(F^{T-t})'} V_{t,\pi}(s,\delta_{m_\mu}) = V_t(s,\delta_{m_\mu}).$$

□

Remark. The assertion is always true for $t = T-1$, if the terminal reward function V'_T and the one step reward function r'_{T-1} do not depend on θ, since then

$$\begin{aligned} V^{\theta}_{T-1,\pi}(s) &= r'_{T-1}(s, f_T(s)) + \mathbb{E}^{\theta}[V'_T(T'_T(s, f_T(s), Y_T))] \\ &= r'_{T-1}(s, f_T(s)) + \theta V'_T(T'_T(s, f_T(s), u)) + (1-\theta)V'_T(T'_T(s, f_T(s), u)), \end{aligned}$$

which is a linear and therefore convex function in θ for all $\pi = f_T \in (F^1)'$.

In the following subsection on examples, we will see that our models do in general not fulfill the strong assumption of Proposition 5.2. We also see, that it is usually not enough to assume convexity of $\theta \mapsto V^{\theta}_{t,\pi^*}(s)$, if π^* is chosen such that

$$\sup_{\pi\in(F^{T-t})'} V_{t,\pi}(s,\delta_{m_\mu}) = V_{t,\pi^*}(s,\delta_{m_\mu}),$$

because the optimal policies in the last equation of the proof differ in general. To this extend, compare also Chapter 5 in Rieder (1987).

To prove the main theorem of this section we need two lemmata. The first one deals with some simple properties of the Bayes operator, while the second one examines the relationship of two properties of the value functions in the case where ϑ is known.

Lemma 5.2. *Let $\mu \in \mathcal{P}(0,1)$ and denote with $\Phi = \Phi_t$ the Bayes operator. Then:*

(i) If $\sigma^2_\mu = 0$, i. e. $\mu \in \{\delta_\theta \,|\, \theta \in [0,1]\}$, then $m_{\Phi(\mu,u)} = m_\mu = m_{\Phi(\mu,d)}$. Otherwise, we have

$$m_{\Phi(\mu,u)} = \frac{m^{(2)}_\mu}{m_\mu}, \qquad m_{\Phi(\mu,d)} = \frac{m_\mu - m^{(2)}_\mu}{1-m_\mu}.$$

(ii) The following equivalences hold:

$$\sigma^2_\mu > 0 \;\Leftrightarrow\; m_{\Phi(\mu,u)} > m_\mu \;\Leftrightarrow\; m_\mu > m_{\Phi(\mu,d)}.$$

(iii) If $\sigma^2_\mu > 0$, i. e. $\mu \notin \{\delta_\theta \,|\, \theta \in [0,1]\}$, then:

$$\frac{m_\mu}{1-m_\mu}(m_{\Phi(\mu,u)} - m_\mu) = m_\mu - m_{\Phi(\mu,d)}.$$

Proof. (i) If $\mu = \delta_\theta$ for some $\theta \in [0,1]$, then $\Phi(\mu, u) = \delta_\theta = \Phi(\mu, d)$ and this case is obvious. Otherwise, $m_\mu \in (0,1)$ and by definition we have for $B \in \mathcal{B}_{[0,1]}$ as in Example 5.1

$$\Phi(\mu, u)(B) = \frac{1}{m_\mu} \int_B \theta \mu(\mathrm{d}\theta),$$

i. e. $\Phi(\mu, u)$ has the μ-density $\frac{\mathrm{id}_{[0,1]}}{m_\mu}$ on the measurable space $([0,1], \mathcal{B}_{[0,1]})$. Analogously, $\Phi(\mu, d)$ has the μ-density $\frac{1-\mathrm{id}_{[0,1]}}{1-m_\mu}$. Therefore,

$$m_{\Phi(\mu,u)} = \int_0^1 \theta\, \Phi(\mu, u)(\mathrm{d}\theta) = \int_0^1 \theta \cdot \frac{\mathrm{id}_{[0,1]}(\theta)}{m_\mu} \mu(\mathrm{d}\theta) = \frac{m_\mu^{(2)}}{m_\mu}$$

and

$$m_{\Phi(\mu,d)} = \int_0^1 \theta\, \Phi(\mu, d)(\mathrm{d}\theta) = \int_0^1 \theta \cdot \frac{1 - \mathrm{id}_{[0,1]}(\theta)}{1 - m_\mu} \mu(\mathrm{d}\theta) = \frac{m_\mu - m_\mu^{(2)}}{1 - m_\mu}.$$

(ii) This follows from the representation in part (i):

$$\begin{aligned} m_{\Phi(\mu,u)} > m_\mu &\Leftrightarrow \frac{m_\mu^{(2)}}{m_\mu} > m_\mu \Leftrightarrow \sigma_\mu^2 = m_\mu^{(2)} - m_\mu^2 > 0 \\ &\Leftrightarrow m_\mu - m_\mu^2 > m_\mu - m_\mu^{(2)} \Leftrightarrow m_\mu > \frac{m_\mu - m_\mu^{(2)}}{1 - m_\mu} \\ &\Leftrightarrow m_\mu > m_{\Phi(\mu,d)}. \end{aligned}$$

(iii) Again, part (i) yields the assertion:

$$\begin{aligned} \frac{m_\mu}{1 - m_\mu}(m_{\Phi(\mu,u)} - m_\mu) &= \frac{m_\mu}{1 - m_\mu}\left(\frac{m_\mu^{(2)}}{m_\mu} - m_\mu\right) \\ &= \frac{m_\mu(1 - m_\mu) - (m_\mu - m_\mu^{(2)})}{1 - m_\mu} \\ &= m_\mu - \frac{m_\mu - m_\mu^{(2)}}{1 - m_\mu} = m_\mu - m_{\Phi(\mu,d)}. \end{aligned}$$

□

Our result is valid for general Bayesian control models with a binomial structure as described above. Recall that the Bayesian model is based on the assumption that for every $\theta \in [0,1]$, we are given a Markov decision process where the reward functions V_T^θ, r_t^θ, $t = 0, 1, \ldots, T-1$ depend on θ. We now make the following assumption.

Assumption 3. *For every* $(s,a) \in D'$*, the functions* $\theta \mapsto V_T^\theta(s)$ *and* $\theta \mapsto r_t^\theta(s,a)$*,* $t = 0, 1, \ldots, T-1$*, are convex on* $[0,1]$*.*

In our examples these functions do not depend on θ, therefore Assumption 3 is trivially fulfilled.

Here comes another lemma. Recall that a function $g : B \to \mathbb{R}$, $B \subset \mathbb{R}^2$, is called *supermodular*, if

$$\begin{aligned} &g(\min\{x_{11}, x_{12}\}, \min\{x_{21}, x_{22}\}) + g(\max\{x_{11}, x_{12}\}, \max\{x_{21}, x_{22}\}) \\ &\geq g(x_{11}, x_{21}) + g(x_{12}, x_{22}) \end{aligned}$$

for all $(x_{11}, x_{21}), (x_{12}, x_{22}) \in B$.

Lemma 5.3. *Consider the model of Section 5.1. If for all* $t \in \{0, 1, \ldots, T\}$ *and* $s \in S'$*,* $a \in A$*, the function*

$$(y, \theta) \;\mapsto\; V_t^\theta(T_t'(s,a,y)), \quad (y,\theta) \in \{u, d\} \times [0,1],$$

is supermodular, then $\theta \mapsto V_t^\theta(s)$ *is convex on* $[0,1]$ *for all* $t \in \{0, 1, \ldots, T\}$ *and* $s \in S'$*.*

Proof. We proceed by backward induction on t. The assertion is clear for $t = T$. Now assume that for fixed $t \in \{1, \ldots, T\}$, the function $V_t^{\cdot}(s)$ is convex for all $s \in S'$ and recall that

$$V_{t-1}^\theta(s) = \sup_{a \in A}\{r_{t-1}^\theta(s,a) + \theta \cdot V_t^\theta(T_t'(s,a,u)) + (1-\theta) \cdot V_t^\theta(T_t'(s,a,d))\}.$$

Define for $s \in S'$, $a \in A$

$$f_{a,s}(\theta) := \theta \cdot V_t^\theta(T_t'(s,a,u)) + (1-\theta) \cdot V_t^\theta(T_t'(s,a,d)), \quad \theta \in [0,1],$$

such that

$$V_{t-1}^\theta(s) = \sup_{a \in A}\{r_{t-1}^\theta(s,a) + f_{a,s}(\theta)\}.$$

Since, by Assumption 3, $\theta \mapsto r_{t-1}^\theta(s,a)$ is convex for every $(s,a) \in D$ and the sum and the supremum of convex functions are again convex, we only have to show that $f_{a,s}$ is convex for all $s \in S'$, $a \in A$. To this extend, introduce the convex (by induction hypothesis) functions $g_u, g_d : [0,1] \to \mathbb{R}$ via

$$g_y(\theta) := V_t^\theta(T_t'(s,a,y)), \quad \theta \in [0,1], y \in \{u, d\},$$

from which we get

$$f_{a,s}(\theta) = \theta \cdot g_u(\theta) + (1-\theta) \cdot g_d(\theta), \quad \theta \in [0,1].$$

By the supermodularity assumption, we have for all $\theta_1, \theta_2 \in [0, 1]$ with $\theta_1 < \theta_2$ that

$$V_t^{\theta_1}(T_t'(s,a,u)) + V_t^{\theta_2}(T_t'(s,a,d)) \leq V_t^{\theta_1}(T_t'(s,a,d)) + V_t^{\theta_2}(T_t'(s,a,u))$$

or equivalently

$$V_t^{\theta_1}(T_t'(s,a,u)) - V_t^{\theta_1}(T_t'(s,a,d)) \leq V_t^{\theta_2}(T_t'(s,a,u)) - V_t^{\theta_2}(T_t'(s,a,d)).$$

We conclude that $g_u - g_d$ is non-decreasing, and we obtain for all $\lambda \in [0, 1]$ and $\theta_1, \theta_2 \in [0, 1]$

$$A(\lambda, \theta_1, \theta_2) := \lambda(1-\lambda)(\theta_1 - \theta_2)(g_u(\theta_1) - g_d(\theta_1) - (g_u(\theta_2) - g_d(\theta_2))) \geq 0.$$

Furthermore, convexity of g_u and g_d yields

$$\begin{aligned}
&f_{a,s}(\lambda\theta_1 + (1-\lambda)\theta_2) \\
&\leq \lambda^2\theta_1 g_u(\theta_1) + \lambda(1-\lambda)\theta_1 g_u(\theta_2) + \lambda(1-\lambda)\theta_2 g_u(\theta_1) \\
&\quad + (1-\lambda)^2\theta_2 g_u(\theta_2) + \lambda^2(1-\theta_1)g_d(\theta_1) + \lambda(1-\lambda)(1-\theta_1)g_d(\theta_2) \\
&\quad + \lambda(1-\lambda)(1-\theta_2)g_d(\theta_1) + (1-\lambda)^2(1-\theta_2)g_d(\theta_2).
\end{aligned}$$

We can write

$$\begin{aligned}
&\lambda f_{a,s}(\theta_1) + (1-\lambda) f_{a,s}(\theta_2) \\
&= \lambda(1-\lambda+\lambda)(\theta_1 g_u(\theta_1) + (1-\theta_1)g_d(\theta_1)) \\
&\quad + (1-\lambda)(1-\lambda+\lambda)(\theta_2 g_u(\theta_2) + (1-\theta_2)g_d(\theta_2)) \\
&= \lambda^2\theta_1 g_u(\theta_1) + \lambda(1-\lambda)\theta_1 g_u(\theta_1) + \lambda(1-\lambda)\theta_2 g_u(\theta_2) \\
&\quad + (1-\lambda)^2\theta_2 g_u(\theta_2) + \lambda^2(1-\theta_1)g_d(\theta_1) + \lambda(1-\lambda)(1-\theta_1)g_d(\theta_1) \\
&\quad + \lambda(1-\lambda)(1-\theta_2)g_d(\theta_2) + (1-\lambda)^2(1-\theta_2)g_d(\theta_2).
\end{aligned}$$

The last two calculations show that

$$\begin{aligned}
&\lambda f_{a,s}(\theta_1) + (1-\lambda) f_{a,s}(\theta_2) - f_{a,s}(\lambda\theta_1 + (1-\lambda)\theta_2) \\
&\geq \lambda(1-\lambda)\theta_1 g_u(\theta_1) + \lambda(1-\lambda)\theta_2 g_u(\theta_2) + \lambda(1-\lambda)(1-\theta_1)g_d(\theta_1) \\
&\quad + \lambda(1-\lambda)(1-\theta_2)g_d(\theta_2) - \lambda(1-\lambda)\theta_1 g_u(\theta_2) - \lambda(1-\lambda)\theta_2 g_u(\theta_1) \\
&\quad - \lambda(1-\lambda)(1-\theta_1)g_d(\theta_2) - \lambda(1-\lambda)(1-\theta_2)g_d(\theta_1) \\
&= \lambda(1-\lambda)(\theta_1 - \theta_2)(g_u(\theta_1) - g_d(\theta_1) - (g_u(\theta_2) - g_d(\theta_2))) = A(\lambda, \theta_1, \theta_2) \geq 0,
\end{aligned}$$

which proves convexity of $f_{a,s}$ and therefore the assertion. □

Recall that if the reward functions depend on $\theta \in [0, 1]$, we set for $(s, \mu, a) \in D$

$$V_T(s,\mu) := \int_{[0,1]} V_T^\theta(s)\,\mu(\mathrm{d}\theta), \quad r_t(s,\mu,a) := \int_{[0,1]} r_t^\theta(s,a)\,\mu(\mathrm{d}\theta), \quad t = 0, 1, \ldots, T-1.$$

We obtain the following comparison result.

Theorem 5.4. *Consider the model of Section 5.1. If for all $t \in \{0, 1, \dots, T\}$ and $s \in S'$, $a \in A$ the function*

$$(y, \theta) \mapsto V_t^\theta(T_t'(s, a, y)), \quad (y, \theta) \in \{u, d\} \times [0, 1],$$

is supermodular, then it holds for all $t \in \{0, 1, \dots, T\}$

$$V_t(s, \mu) \geq V_t(s, \delta_{m_\mu}), \quad (s, \mu) \in S.$$

Proof. The proof is again by backward induction on t. For $t = T$ we have to show that

$$V_T(s, \mu) = \int_{[0,1]} V_T^\theta(s)\, \mu(\mathrm{d}\theta) \geq V_T^{m_\mu}(s) = V_T(s, \delta_{m_\mu}), \quad (s, \mu) \in S.$$

But this is again a simple consequence of Jensen's inequality.
Now assume that the assertion holds for fixed $t \in \{1, \dots, T\}$ and let $(s, \mu) \in S$, where $\mu \in \mathcal{P}(0, 1)$ with $\mu \notin \{\delta_\theta \mid \theta \in [0, 1]\}$. Note that then $\Phi_t(\mu, y) \in \mathcal{P}(0, 1)$ for $y \in \{u, d\}$. Furthermore, by definition of the transition kernel we have in the same way as in (5.7)

$$Q_t(s, \mu, a; \{x_u\} \times \{\Phi_t(\mu, u)\}) = m_\mu = 1 - Q_t(s, \mu, a; \{x_d\} \times \{\Phi_t(\mu, d)\}). \quad (5.14)$$

Again, we obtain a two-point distribution on

$$G := \{(x_u, \Phi_t(\mu, u)), (x_d, \Phi_t(\mu, d))\}$$

Additionally, we can treat the one-step reward function in the same way as V_T above and obtain for $a \in A$

$$r_{t-1}(s, \mu, a) = \int_{[0,1]} r_{t-1}^\theta(s, a)\, \mu(\mathrm{d}\theta) \geq r_{t-1}^{m_\mu}(s, a) = r_{t-1}(s, \delta_{m_\mu}, a),$$

since, by model assumption, $r_{t-1}^{\cdot}(s, a)$ is convex on $[0, 1]$. Consequently, the induction hypothesis (I.H.) yields with $\mu_u := \Phi_t(\mu, u)$, $\mu_d := \Phi_t(\mu, d)$ and with $m_u := m_{\Phi_t(\mu,u)}$, $m_d := m_{\Phi_t(\mu,d)}$

$$\begin{aligned}
&V_{t-1}(s, \mu) \\
&= \sup_{a \in A} \Big\{ r_{t-1}(s, \mu, a) + \int_G V_t(x) Q_t(s, \mu, a; \mathrm{d}x) \Big\} \\
&\overset{(5.14)}{=} \sup_{a \in A} \Big\{ r_{t-1}(s, \mu, a) + m_\mu \cdot V_t(T_t'(s, a, u), \mu_u) + (1 - m_\mu) \cdot V_t(T_t'(s, a, d), \mu_d) \Big\} \\
&\overset{\text{I.H.}}{\geq} \sup_{a \in A} \Big\{ r_{t-1}(s, \delta_{m_\mu}, a) + m_\mu \cdot V_t(T_t'(s, a, u), \delta_{m_u}) + (1 - m_\mu) \cdot V_t(T_t'(s, a, d), \delta_{m_d}) \Big\}.
\end{aligned}$$

The introduced lemmata provide now the result as follows. From Lemma 5.2(ii) we have $m_d < m_\mu < m_u$ and part (iii) yields

$$\frac{m_\mu}{1-m_\mu} = \frac{m_\mu - m_d}{m_u - m_\mu}. \tag{5.15}$$

Furthermore, by assumption and Lemma 5.3, $V_t(s', \delta_\cdot)$ is convex on $[0,1]$ for every $s' \in S'$ and therefore

$$\frac{V_t(s', \delta_{m_\mu}) - V_t(s', \delta_{m_d})}{m_\mu - m_d} \le \frac{V_t(s', \delta_{m_u}) - V_t(s', \delta_{m_\mu})}{m_u - m_\mu} \tag{5.16}$$

holds. Again, we obtain with $s' = T_t'(s,a,u)$, $a \in A$, by the supermodularity assumption

$$\begin{aligned}
&V_t(T_t'(s,a,d), \delta_{m_\mu}) - V_t(T_t'(s,a,d), \delta_{m_d}) \\
&\le V_t(T_t'(s,a,u), \delta_{m_\mu}) - V_t(T_t'(s,a,u), \delta_{m_d}) \\
&\overset{(5.16)}{\le} \frac{m_\mu - m_d}{m_u - m_\mu}\big(V_t(T_t'(s,a,u), \delta_{m_u}) - V_t(T_t'(s,a,u), \delta_{m_\mu})\big) \\
&\overset{(5.15)}{=} \frac{m_\mu}{1-m_\mu}\big(V_t(T_t'(s,a,u), \delta_{m_u}) - V_t(T_t'(s,a,u), \delta_{m_\mu})\big),
\end{aligned}$$

which is equivalent to

$$\begin{aligned}
&m_\mu \cdot V_t(T_t'(s,a,u), \delta_{m_u}) + (1-m_\mu) \cdot V_t(T_t'(s,a,d), \delta_{m_d}) \\
&\ge m_\mu \cdot V_t(T_t'(s,a,u), \delta_{m_\mu}) + (1-m_\mu) \cdot V_t(T_t'(s,a,d), \delta_{m_\mu}).
\end{aligned}$$

Combining this with the above inequality for V_{t-1}, we get

$$\begin{aligned}
&V_{t-1}(s,\mu) \\
&\ge \sup_{a\in A}\big\{r_{t-1}(s, \delta_{m_\mu}, a) + m_\mu \cdot V_t(T_t'(s,a,u), \delta_{m_u}) + (1-m_\mu) \cdot V_t(T_t'(s,a,d), \delta_{m_d})\big\} \\
&\ge \sup_{a\in A}\big\{r_{t-1}(s, \delta_{m_\mu}, a) + m_\mu \cdot V_t(T_t'(s,a,u), \delta_{m_\mu}) + (1-m_\mu) \cdot V_t(T_t'(s,a,d), \delta_{m_\mu})\big\} \\
&= V_{t-1}(s, \delta_{m_\mu}),
\end{aligned}$$

thus completing the proof. □

Remark. If the terminal reward function V_T and the one-step reward function r_{T-1} do not depend on θ we have equality of the value functions for $t = T-1$:

$$\begin{aligned}
V_{T-1}(s,\mu) &= \sup_{a\in A}\{r_{T-1}'(s,a) + \int_E V_T'(T_T'(s,a,y))Q_T(s,\mu,a;dy)\} \\
&= \sup_{a\in A}\{r_{T-1}'(s,a) + m_\mu V_T'(T_T'(s,a,u)) + (1-m_\mu)V_T'(T_T'(s,a,d))\} \\
&= V_{T-1}(s, \delta_{m_\mu}).
\end{aligned}$$

The result can, of course, be used to compare our two dynamic risk measures. Consider the reward functions as introduced in the previous chapter (or the previous section).

Corollary 5.1. *Let the assumption of Theorem 5.4 be fulfilled. Take a wealth process* $(W_t)_{t=0,1,\dots,T}$ *and the sequence of posterior distributions defined using the prior distribution* $\mu_0 \in \mathcal{P}(0,1)$ *and* $\mu_{t+1} := \Phi_t(\mu_t, Y_t)$, $t = 0, 1, \dots, T-1$. *Then*

$$\rho_t^{\mathrm{PR}, m_{\mu_t}}(I) \geq \rho_t^{\mathrm{B}, \mu_0}(I), \quad I \in \mathcal{X}^M.$$

In particular, if we use the Beta distributions as a conjugate family for the binomial distribution, we set $\mu_0 := \mathcal{U}(0,1)$ *and obtain*

$$\rho_0^{\mathrm{PR}, \frac{1}{2}}(I) \geq \rho_0^{\mathrm{B}, \mathcal{U}(0,1)}(I), \quad I \in \mathcal{X}^M.$$

Proof. This follows directly from Theorem 5.4 with $v_t(w) := -\frac{q_t}{c_t} w^-$, $w \in S_1$:

$$\begin{aligned}\rho_t^{\mathrm{PR}, m_{\mu_t}}(I) &= -\frac{q_t}{c_t} W_t^- - \frac{1}{c_t} V_t^{m_{\mu_t}}(W_t, Z_t) = -\frac{q_t}{c_t} W_t^- - \frac{1}{c_t} V_t(W_t, Z_t, \delta_{m_{\mu_t}}) \\ &\geq -\frac{q_t}{c_t} W_t^- - \frac{1}{c_t} V_t(W_t, Z_t, \mu_t) = \rho_t^{\mathrm{B}, \mu_0}(I).\end{aligned}$$

□

Economic interpretation for t = 0. We see that if the supermodularity assumption is fulfilled the dynamic risk measure ρ^{PR} is more conservative than ρ^{B} since it assigns a higher risk to every process $I \in \mathcal{X}^{\mathrm{M}}$. In the next section we will see that this is also the case in the Artzner–game although the supermodularity assumption is not fulfilled there.

The interpretation of this fact is as follows. Recall that we assume that the probability distribution of ϑ is unknown and, in order to deal with this fact, we choose an initial distribution as an estimation for this distribution. Therefore, when the parameter is known with realization $\theta_0 \in (0,1)$ using the dynamic risk measure ρ^{PR} can also be interpreted as using δ_{θ_0} as an initial distribution, formally,

$$\rho_0^{\mathrm{PR}, \theta_0}(I) = \rho_0^{\mathrm{B}, \delta_{\theta_0}}(I), \quad I \in \mathcal{X}^{\mathrm{M}}.$$

As we have seen, in contrast to using a general initial distribution μ_0, the current estimation of $\mathcal{L}(\vartheta)$ remains constant if we start with δ_{θ_0}, i.e. the information over time, namely the realizations of the generating process $(Y_t)_{t=1,\dots,T}$, are not taken into account when calculating the dynamic risk measure. On the other hand, this is the case if the initial distribution is μ_0. Therefore, we are in fact able to diminish our risk at $t = 0$ by using the information revealed over time. In this way, the risk of the process I is assessed more adequately.

To conclude this subsection, we further investigate the relationship of convexity of $V_t^{\cdot}(s)$ and the result of the theorem.

Proposition 5.3. *Let $t \in \{0, 1, \ldots, T\}$, $s \in S'$ and assume that the inequality $V_t(s, \delta_{m_\mu}) \leq V_t(s, \mu)$ holds for all $\mu \in \mathcal{P}(0, 1)$. Then $\theta \mapsto V_t^\theta(s)$ is convex on $[0, 1]$.*

Proof. It is well known (compare e. g. Chapter 5 in Rieder (1987))) that $V_t(s, \cdot)$ is convex on $\mathcal{P}(0, 1)$. This can be seen by taking $\mu_1, \mu_2 \in \mathcal{P}(0, 1)$, $\lambda \in [0, 1]$ and

$$\begin{aligned}
&V_t(s, \lambda\mu_1 + (1-\lambda)\mu_2) \\
&= \sup_{\pi \in (F^{T-t})'} \int_\Theta V_{t,\pi}^\theta(s)\,(\lambda\mu_1 + (1-\lambda)\mu_2)(\mathrm{d}\theta) \\
&= \sup_{\pi \in (F^{T-t})'} \Big\{\lambda \int_\Theta V_{t,\pi}^\theta(s)\,\mu_1(\mathrm{d}\theta) + (1-\lambda)\int_\Theta V_{t,\pi}^\theta(s)\,\mu_2(\mathrm{d}\theta)\Big\} \\
&\leq \lambda \sup_{\pi \in (F^{T-t})'} \int_\Theta V_{t,\pi}^\theta(s)\,\mu_1(\mathrm{d}\theta) + (1-\lambda) \sup_{\pi \in (F^{T-t})'} \int_\Theta V_{t,\pi}^\theta(s)\,\mu_2(\mathrm{d}\theta) \\
&= \lambda V_t(s, \mu_1) + (1-\lambda)V_t(s, \mu_2).
\end{aligned}$$

Now take $\theta_1, \theta_2 \in [0, 1]$, $\lambda \in [0, 1]$ and note that $m_{\lambda\delta_{\theta_1} + (1-\lambda)\delta_{\theta_2}} = \lambda\theta_1 + (1-\lambda)\theta_2$. The assumption and convexity of $V_t(\varepsilon, \cdot)$ then yield the assertion:

$$\begin{aligned}
V_t^{\lambda\theta_1 + (1-\lambda)\theta_2}(s) &= V_t(s, \delta_{\lambda\theta_1 + (1-\lambda)\theta_2}) \leq V_t(s, \lambda\delta_{\theta_1} + (1-\lambda)\delta_{\theta_2}) \\
&\leq \lambda V_t(s, \delta_{\theta_1}) + (1-\lambda)V_t(s, \delta_{\theta_2}) = \lambda V_t^{\theta_1}(s) + (1-\lambda)V_t^{\theta_2}(s).
\end{aligned}$$

□

5.4.2 Examples

In this subsection we treat some examples, namely the coin tossing game proposed by Artzner and the standard Cox Ross Rubinstein model (CRRM).

Examples and counterexamples in the Artzner–game Consider the Artzner example. We have $E = \{0, 1\}$ and $S' = \mathbb{R} \times \mathbb{N}_0$, whereas because of $T = 3$ also $S' = \mathbb{R} \times \{0, 1, 2, 3\}$ can be used. But this makes no difference for our investigations. Take the process

$$I_1 = I_2 = 0,\ I_3 = \mathbb{1}_{\{Z_2 + Y_3 \geq 2\}}.$$

First, let us check that the inductive proof of Theorem 5.4 does not work if we only assume convexity of $V_t(s)$, $s \in S'$. Consider $s = (w, z) \in S'$ and define with $\theta \in [0, 1]$

$$p_z(\theta) := P(Y_3 \geq 2 - z \mid \vartheta = \theta) = \begin{cases} 0 & , \quad z = 0, \\ \theta & , \quad z = 1, \\ 1 & , \quad z \geq 2. \end{cases}$$

Then

$$\begin{aligned}
V_2(s,\delta_\theta) &= c_3 w^+ - q_2 w^- \\
&\quad + c_4 \mathbb{E}\left[I_3 \,|\, Z_2 = z, \vartheta = \theta\right] - (c_3 - c_4)\mathrm{AV@R}_{\gamma_3}(I_3 \,|\, Z_2 = z, \vartheta = \theta) \\
&= c_3 w^+ - q_2 w^- \\
&\quad + c_4 \mathbb{E}\left[\mathbb{1}_{\{Y_3 \geq 2-z\}} \,|\, \vartheta = \theta\right] - (c_3 - c_4)\mathrm{AV@R}_{\gamma_3}(\mathbb{1}_{\{Y_3 \geq 2-z\}} \,|\, \vartheta = \theta) \\
&= c_3 w^+ - q_2 w^- + c_4 p_z(\theta) + (c_3 - c_4)\frac{p_z(\theta) - \gamma_3}{1 - \gamma_3}\mathbb{1}_{[0,p_z(\theta)]}(\gamma_3),
\end{aligned}$$

which is obviously convex in θ for every $s = (w, z) \in S'$. But the main assumption of Theorem 5.4 does not hold, as can be seen as follows. Take an arbitrary $a \in A$ and set $z = 1$, $w = 0$. Then we have

$$\begin{aligned}
T_2'(w,z,a,u) &= (w^+ + h_2(z,u) - a, s_2 + u) = (-a, 2) \\
T_2'(w,z,a,d) &= (w^+ + h_2(z,d) - a, s_2 + d) = (-a, 1).
\end{aligned}$$

Taking $\theta_1, \theta_2 \in (0,1)$ with $\theta_1 < \theta_2 < \gamma_3$ we obtain with the above calculation and because $p_{\theta_i}(1) = \theta_i$, $i = 1, 2$,

$$V_2(T_2'(w,z,a,d),\delta_{\theta_2}) - V_2(T_2'(w,z,a,d),\delta_{\theta_1}) = c_4 p_{\theta_2}(1) - c_4 p_{\theta_2}(1) = c_4(\theta_2 - \theta_1) > 0$$

and, since $p_{\theta_i}(2) = 1$, $i = 1, 2$,

$$V_2(T_2'(w,z,a,u),\delta_{\theta_2}) - V_2(T_2'(w,z,a,u),\delta_{\theta_1}) = c_4 p_{\theta_2}(2) - c_4 p_{\theta_2}(2) = 0.$$

We first conclude that even in the case $T = t = 1$, the reverse implication in Lemma 5.3 is not true. Also note that for $T = 3$, a comparison result for V_2 is provided by the remark after Theorem 5.4.

Furthermore, we will now see that convexity of $V_2^{\cdot}(s)$ is not enough to imply the assertion of Theorem 5.4, i. e. we will find a pair $(s, \mu) \in S' \times \mathcal{P}(0,1)$ such that

$$V_1(s,\mu) < V_1(s,\delta_{m\mu}).$$

We can assume without loss of generality that $w = 0$ and have to consider the two cases $z = 0$ and $z = 1$, which are just the possible realizations of $Z_1 = Y_1$.

First, we choose $s = (0,1)$ and consequently $\mu = \mathrm{Beta}(2,1)$, such that $m_\mu = \frac{2}{3}$. Furthermore let $\gamma_3 > \frac{2}{3}$, e. g. $\gamma_3 = 0.95$. In this case, we have $\mathrm{AV@R}^{\theta}_{\gamma_3}(Y_3) = 0$ for all $\theta \leq 0.95$. Subsection 5.3.1 yields

$$\begin{aligned}
V_1(0,1,\delta_{\frac{2}{3}}) &= -c_3 \cdot \mathbb{E}^{\frac{2}{3}}\left[\rho^{(3)}_{\frac{2}{3}}(\mathbb{1}_{\{Z_2+Y_3 \geq 2\}} \,|\, Z_2 = 1 + Y_2)\right] \\
&= -c_3 \cdot \left(\frac{2}{3}\rho^{(3)}_{\frac{2}{3}}(\mathbb{1}_{\{Y_3 \geq 0\}}) + \frac{1}{3}\rho^{(3)}_{\frac{2}{3}}(\mathbb{1}_{\{Y_3 \geq 1\}})\right) \\
&= -c_3 \cdot \left(\frac{2}{3} \cdot (-1) - \frac{1}{3} \cdot \frac{2}{3} \cdot \lambda_3\right) = \frac{2}{9}c_4 + \frac{2}{3}c_3,
\end{aligned}$$

while Proposition 5.1 implies

$$\begin{aligned} V_1\big(0,1,\mathrm{Beta}(2,1)\big) &= -c_3 \cdot \mathbb{E}_{1,2,1}\Big[\rho^{(3)}_{1,2,1}\big(\mathbb{1}_{\{Z_2+Y_3\ge 2\}} \,\big|\, Y_2, Z_1 = 1\big)\Big] \\ &= -c_3 \cdot \mathbb{E}\Big[\rho^{(3)}\Big(\mathbb{1}_{\{Y_2+Y_3\ge 1\}} \,\Big|\, Y_2, \vartheta = \frac{2+\mathbb{1}_{\{1\}}(Y_2)}{4}\Big) \,\Big|\, \vartheta = \frac{2}{3}\Big] \\ &= -c_3 \cdot \Big(\frac{2}{3}\cdot \rho^{(3)}_{\frac{3}{4}}\big(\mathbb{1}_{\{Y_3\ge 0\}}\big) + \frac{1}{3}\cdot \rho^{(3)}_{\frac{1}{2}}\big(\mathbb{1}_{\{Y_3\ge 1\}}\big)\Big) \\ &= -c_3 \cdot \Big(\frac{2}{3}\cdot(-1) - \frac{1}{3}\cdot\frac{1}{2}\lambda_3\Big) = \frac{1}{6}c_4 + \frac{2}{3}c_3 < V_1\big(0,1,\delta_{\frac{2}{3}}\big). \end{aligned}$$

Similar calculations show that for $s = (0,0)$ and thus $\mu = \mathrm{Beta}(1,2)$ the reverse inequality holds:

$$V_1\big(0,0,\delta_{\frac{1}{3}}\big) = \frac{1}{9}c_4 < \frac{1}{6}c_4 = V_1\big(0,0,\mathrm{Beta}(1,2)\big).$$

So, indeed assuming convexity in Theorem 5.4 is not enough to ensure the desired result. We rather have to make the stronger (by Lemma 5.3) assumption on $D_{t,s,a}$.

So the theorem does not provide a comparison result for the dynamic risk measures when considering the Artzner example. But since the problem is very simple, all quantities can easily be computed and calculated. We have already done so and compared the value functions for $t = 1$ and $t = 2$. Finally, we find for the value functions with $t = 0$, $s = (0,0)$ and $\mu = \mathrm{Beta}(1,1) = \mathcal{U}(0,1)$, therefore $m_\mu = \frac{1}{2}$,

$$\begin{aligned} &V_0\big(0,0,\delta_{\frac{1}{2}}\big) \\ &= -c_3 \cdot \mathbb{E}^{\frac{1}{2}}\big[\rho^{(3)}_{\frac{1}{2}}\big(\mathbb{1}_{\{Z_2+Y_3\ge 2\}} \,|\, Z_2\big)\big] \\ &= -c_3 \cdot \Big(\frac{1}{4}\rho^{(3)}_{\frac{1}{2}}\big(\mathbb{1}_{\{Y_3\ge 0\}}\big) + \frac{1}{2}\rho^{(3)}_{\frac{1}{2}}\big(\mathbb{1}_{\{Y_3\ge 1\}}\big) + \frac{1}{2}\rho^{(3)}_{\frac{1}{2}}\big(\mathbb{1}_{\{Y_3\ge 2\}}\big)\Big) \\ &= -c_3 \cdot \Big(-\frac{1}{4} - \frac{1}{2}\cdot\frac{1}{2}\cdot\lambda_3 + \frac{1}{2}\cdot 0\Big) = \frac{1}{4}c_3 + \frac{1}{4}c_4 \end{aligned}$$

and

$$\begin{aligned} &V_0\big(0,0,\mathcal{U}(0,1)\big) \\ &= -c_3 \cdot \mathbb{E}_{0,1,1}\big[\rho^{(3)}_{0,1,1}\big(\mathbb{1}_{\{Z_2+Y_3\ge 2\}} \,|\, Y_2, Y_1\big)\big] \\ &= -c_3 \cdot \Big(\frac{1}{2}\cdot\frac{2}{3}\cdot\rho^{(3)}_{\frac{3}{4}}\big(\mathbb{1}_{\{Y_3\ge 0\}}\big) + 2\cdot\frac{1}{2}\cdot\frac{1}{3}\cdot\rho^{(3)}_{\frac{1}{2}}\big(\mathbb{1}_{\{Y_3\ge 1\}}\big) + \frac{1}{2}\cdot\frac{2}{3}\cdot\rho^{(3)}_{\frac{1}{4}}\big(\mathbb{1}_{\{Y_3\ge 2\}}\big)\Big)\Big) \\ &= -c_3 \cdot \Big(-\frac{1}{3} - \frac{1}{3}\cdot\frac{1}{2}\cdot\lambda_3 + 0\Big) = \frac{1}{3}c_3 + \frac{1}{6}c_4 > V_0\big(0,0,\delta_{\frac{1}{2}}\big), \end{aligned}$$

since $c_4 < c_3$. We conclude that the risk measures have the reverse order:

$$\rho_0^{\mathrm{PR},\frac{1}{2}}(I) > \rho_0^{\mathrm{B},\mathcal{U}(0,1)}(I).$$

$\rho_0^{\mathrm{PR},\frac{1}{2}}(I^{(1)})$	-0.4325		
$\rho_0^{\mathrm{B}\mathcal{U}(0,1)}(I^{(1)})$	-0.4325		
$\rho_0^{\mathrm{PR},\frac{1}{2}}(I^{(2)})$	-0.4419		
$\rho_0^{\mathrm{B}\mathcal{U}(0,1)}(I^{(2)})$	-0.445		
	$Y_1 = 1$		$Y_1 = 0$
$\rho_1^{\mathrm{PR},\frac{1+Y_1}{3}}(I^{(1)})$	-0.5766		-0.2883
$\rho_1^{\mathrm{B}\mathcal{U}(0,1)}(I^{(1)})$	-0.5766		-0.2883
$\rho_1^{\mathrm{PR},\frac{1+Y_1}{3}}(I^{(2)})$	-0.7939		-0.0961
$\rho_1^{\mathrm{B}\mathcal{U}(0,1)}(I^{(2)})$	-0.7458		-0.1442
	$Y_1 + Y_2 = 2$	$Y_1 + Y_2 = 1$	$Y_1 + Y_2 = 0$
$\rho_2^{\mathrm{PR},\frac{1+Y_1+Y_2}{4}}(I^{(1)})$	-0.6828	-0.4552	-0.2276
$\rho_2^{\mathrm{B}\mathcal{U}(0,1)}(I^{(1)})$	-0.6828	-0.4552	-0.2276
$\rho_2^{\mathrm{PR},\frac{1+Y_1+Y_2}{4}}(I^{(2)})$	-0.95	-0.4552	0
$\rho_2^{\mathrm{B}\mathcal{U}(0,1)}(I^{(2)})$	-0.95	-0.4552	0

Table 5.1: The dynamic risk measures in the Artzner–game

In Table 5.1 we give some numerical values of the risk measures, where the variables $c_k = 0.95^{k-1}$, $k = 1, 2, 3$, $c_4 = 0.93^2$ and $q_k = 1.2 \cdot c_k$, $k = 1, 2, 2$, are chosen analogously to Pflug and Ruszczyński (2001). In that work, the values for $\rho_0^{\mathrm{PR},\frac{1}{2}}(I^{(1)})$ and $\rho_0^{\mathrm{PR},\frac{1}{2}}(I^{(2)})$ can already be found.

Examples and counterexamples in the CRRM Now, let us treat the CRRM. First, we want to show that the assumption of Proposition 5.2 is too strong. It is easily seen that $\theta \mapsto V^{\theta}_{T-1,\pi}(s)$ is linear, therefore convex, for every $s \in S'$ and $\pi = f_T \in (F^1)'$. So let us take $t = T - 2$. Furthermore, choose $s = (0, 1) \in S'$ and the admissible policy $\pi = (f_{T-1}, f_T) \in (F^2)'$ defined through

$$f_{T-1}(w, z) := -1, \quad f_T(w, z) := w^+ + \mathbb{1}_{\{z=u\}} u(d - 1), \quad (w, z) \in S'.$$

Starting with $(W_{T-2}, Z_{T-2}) = (w, z) = (0, 1)$ and using the policy π from above yields $W_{T-1} = w^+ + z(Y_{T-1} - 1) - f_{T-1}(w, z) = Y_{T-1}$, so $W^+_{T-1} = Y_{T-1}$ and $W^-_{T-1} = 0$. Furthermore,

$$\begin{aligned} W_T &= W^+_{T-1} + Z_{T-2} \cdot Y_{T-1}(Y_T - 1) - f_T(W_{T-1}, Z_{T-1}) \\ &= Y_{T-1}(Y_T - 1) - \mathbb{1}_{\{Y_{T-1}=u\}} u(d - 1). \end{aligned}$$

With the reward functions used in Section 5.3 we obtain

$$\begin{aligned}
&V^\theta_{T-2,\pi}(0,1)\\
&= \mathbb{E}\,[r_{T-2}(X_{T-2}, f_{T-1}(X_{T-2})) + r_{T-1}(X_{T-1}, f_T(X_{T-1}))\\
&\qquad\qquad\qquad\qquad\qquad\qquad + V_T(X_T) \mid X_{T-2} = (0,1,\delta_\theta)]\\
&= 0 + \mathbb{E}^\theta[-q_{T-1}W^-_{T-1} + c_T f_T(X_{T-1}) + c_{T+1}W^+_T - q_T W^-_T]\\
&= \mathbb{E}^\theta[c_T(Y_{T-1} + \mathbb{1}_{\{Y_{T-1}=u\}}u(d-1)) + c_{T+1}(Y_{T-1}(Y_T-1) - \mathbb{1}_{\{Y_{T-1}=u\}}u(d-1))^+\\
&\quad - q_T(Y_{T-1}(Y_T-1) - \mathbb{1}_{\{Y_{T-1}=u\}}u(d-1))^-]\\
&= c_T(\theta ud + (1-\theta)d) + \theta^2(c_{T+1}[u^2-ud]^+ - q_T[u^2-ud]^-) + \theta(1-\theta)\cdot 0\\
&\quad + (1-\theta)\theta(c_{T+1}[d(u-1)]^+ - q_T[d(u-1)]^-)\\
&\quad + (1-\theta)^2(c_{T+1}[d(d-1)]^+ - q_T[d(d-1)]^-)\\
&= c_T(\theta ud + (1-\theta)d) + \theta^2 c_{T+1}u(u-d) + (\theta-\theta^2)c_{T+1}d(u-1)\\
&\quad - (1-2\theta+\theta^2)q_T d(1-d).
\end{aligned}$$

This yields for $u = 1$

$$\begin{aligned}
\frac{\partial^2 V^\theta_{T-2,\pi}(s)}{\partial\theta^2} &= 2c_{T+1}u(u-d) - 2c_{T+1}d(u-1) - 2q_T d(1-d)\\
&= 2c_{T+1}(1-d) - 2q_T d(1-d) = 2(1-d)(c_{T+1} - q_T d),
\end{aligned}$$

which is less than zero, if $\frac{c_{T+1}}{q_T} < d$, which is clearly a possible choice. So in this example, $\theta \mapsto V^\theta_{T-2,\pi}(s)$ is concave and consequently Proposition 5.2 cannot be applied to obtain comparison results in the CRRM with our reward functions. But as we will see now, the assumption of Theorem 5.4 is fulfilled.

Note that the state space is $S' = \mathbb{R} \times \mathbb{R}^*_+$, therefore $Z_t > 0$ for all $t = 0, 1, \ldots, T$. Let us first calculate the value functions $V^\theta_t(s)$ for $s \in S'$ and $\theta \in [0,1]$ at time $t \in \{1, \ldots, T\}$. Because of (conditional) coherence of the risk measure $\rho^{(k)}_\theta$ for every $k = t+1, \ldots, T$ Theorem 4.2 yields

$$\begin{aligned}
&V^\theta_t(w,z) - c_{t+1}w^+ + q_t w^-\\
&\quad = \sum_{k=t+1}^T c_k \cdot \mathbb{E}^\theta[-\rho^{(k)}_\theta(Z_{k-1}(Y_k - 1) \mid Z_{k-1}) \mid Z_t = z]\\
&\quad = \sum_{k=t+1}^T c_k \cdot \mathbb{E}^\theta[Z_{k-1} \mid Z_t = z] \cdot (-\rho^{(k)}_\theta(Y_k) - 1)\\
&\quad = \sum_{k=t+1}^T c_k \cdot z \cdot \mathbb{E}^\theta[Y_1]^{k-(t+1)} \cdot (\lambda_k \mathbb{E}^\theta[Y_k] - (1-\lambda_k)\mathrm{AV@R}^\theta_{\gamma_k}(Y_k) - 1)
\end{aligned}$$

$$= z \cdot \sum_{k=t+1}^{T} c_k \cdot \mathbb{E}^{\theta}[Y_1]^{k-(t+1)} \times \Big(\lambda_k\theta(u-d) + d - 1 + (1-\lambda_k)\mathbb{1}_{[0,\theta]}(\gamma_k)\frac{\theta-\gamma_k}{1-\gamma_k}(u-d)\Big).$$

Now, let $(s,a) = (w,z,a) \in S' \times A$. Recall that $T_t'(s,a,y) = (w^+ + z(y-1) - a, zy)$, $y \in \{u,d\}$. We obtain

$$\begin{aligned}
D_{t,s,a}(\theta) &= V_t^{\theta}(T_t'(s,a,u)) - V_t^{\theta}(T_t'(s,a,d)) \\
&= c_{t+1}(w^+ + z(u-1) - a)^+ - q_t(w^+ + z(u-1) - a)^- \\
&\quad - c_{t+1}(w^+ + z(d-1) - a)^+ + q_t(w^+ + z(d-1) - a)^- \\
&\quad + z(u-d) \cdot \sum_{k=t+1}^{T} c_k \cdot \mathbb{E}^{\theta}[Y_1]^{k-(t+1)} \times \\
&\qquad \Big(\lambda_k\theta(u-d) + d - 1 + (1-\lambda_k)\mathbb{1}_{[0,\theta]}(\gamma_k)\frac{\theta-\gamma_k}{1-\gamma_k}(u-d)\Big) \\
&= c_{t+1}(w^+ + z(u-1) - a)^+ - q_t(w^+ + z(u-1) - a)^- \\
&\quad - c_{t+1}(w^+ + z(d-1) - a)^+ + q_t(w^+ + z(d-1) - a)^- \\
&\quad + z(u-d) \cdot \sum_{k=t+1}^{T} c_k \cdot \mathbb{E}^{\theta}[Y_1]^{k-(t+1)}(\lambda_k\theta(u-d) + d - 1) \\
&\quad + z(u-d) \cdot \sum_{k=t+1}^{T} c_k \cdot \mathbb{E}^{\theta}[Y_1]^{k-(t+1)}(1-\lambda_k)\mathbb{1}_{[0,\theta]}(\gamma_k)\frac{\theta-\gamma_k}{1-\gamma_k}(u-d).
\end{aligned}$$

We have to show that this term is non–decreasing in θ. To this extend, we only have to consider the two sums. Since $\mathbb{E}^{\theta}[Y_1] = \theta(u-d)+d$, the last one is the sum of non–negative products of non–decreasing functions, therefore clearly non–decreasing. The first sum is just the value function $V_t^{\theta}(0, z(u-d))$, if $\theta \leq \min\{\gamma_{t+1}, \ldots, \gamma_T\}$. So we only have to show that $V_t^{\theta}(s)$ is non–decreasing in θ on this interval for every $s \in S'$. This can be seen as follows. The case $t = T-1$ is obvious, so assume $t \leq T-2$. Furthermore, without loss of generality, we can take $(w,z) = (0,1)$. Let us write the value function as

$$\begin{aligned}
&V_t^{\theta}(w,z) \\
&= c_{t+1}(\lambda_{t+1}\theta(u-d) + d - 1) + c_{t+2}(\theta(u-d) + d)(\lambda_{t+2}\theta(u-d) + d - 1) \\
&\quad + \sum_{k=t+3}^{T} c_k \cdot \mathbb{E}^{\theta}[Y_1]^{k-(t+1)}(\lambda_k\theta(u-d) + d) - \sum_{k=t+3}^{T} c_k \cdot \mathbb{E}^{\theta}[Y_1]^{k-(t+1)}.
\end{aligned}$$

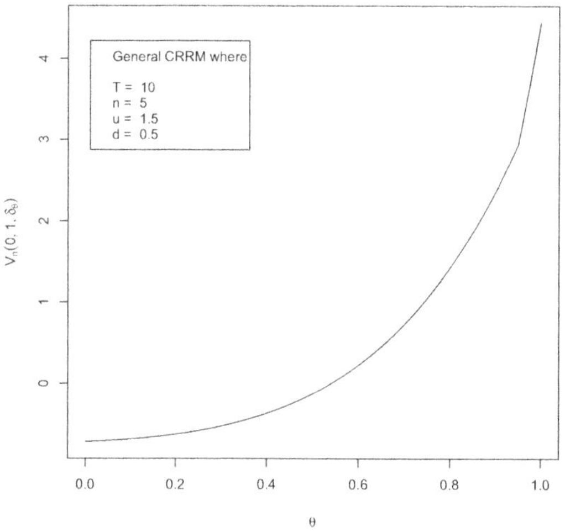

Figure 5.1: The value function in the CRRM

We obtain

$$
\begin{aligned}
&\frac{\partial V_t^\theta(w,z)}{\partial\theta}\\
&= c_{t+2}(u-d) + \underbrace{2c_{t+3}(u-d)^2\theta + c_{t+3}d(u-d) + \overbrace{c_{t+2}}^{\geq c_{t+3}}(u-d)d}_{\geq 2c_{t+3}(u-d)(\theta(u-d)+d)} - c_{t+2}(u-d)\\
&\quad + \sum_{k=t+3}^{T} c_k(u-d)\mathbb{E}^{\theta}[Y_1]^{k-(t+1)-1}\Big((k-(t+1))(\lambda_k\theta(u-d) + \underbrace{d}_{\geq\lambda_k d}) + \lambda_k\mathbb{E}^{\theta}[Y_1]\Big)\\
&\quad - 2c_{t+3}(u-d)\mathbb{E}^{\theta}[Y_1] - \sum_{k=t+4}^{T} c_k(u-d)\mathbb{E}^{\theta}[Y_1]^{k-(t+1)-1}(k-(t+1))\\
&\geq \sum_{k=t+3}^{T} \overbrace{c_k\lambda_k}^{=c_{k+1}}(u-d)\mathbb{E}^{\theta}[Y_1]^{k-(t+1)-1}(k-t)\cdot\mathbb{E}^{\theta}[Y_1]\\
&\quad - \sum_{k=t+3}^{T-1} c_{k+1}(u-d)\mathbb{E}^{\theta}[Y_1]^{k-(t+1)}(k-t)\\
&= c_{T+1}(u-d)\mathbb{E}^{\theta}[Y_1]^{T-t-1}(T-t) \geq 0.
\end{aligned}
$$

We conclude that Theorem 5.4 respectively Corollary 5.1 can be applied to the CRRM when using the reward functions that result from the risk measure ρ^{PR}. To visualize the behaviour of the value function, we provide a simple graph in Figure 5.1. We set $c_k = 0.95^{k-1}$, $k = 1, \dots, T$, $c_{T+1} = 0.93^{T-1}$ and $\gamma_k = 0.95$, $k = 1, \dots, T$. Notice the point of non–differentiability at $\theta = 0.95 = \gamma_T$.

A Absolutely continuous probability measures

Let $(\Omega, \mathcal{F})$ be a measurable space and $\mathbb{P}$ a probability measure on it. Denote with $L^p(\Omega, \mathcal{F}, \mathbb{P})$ for $p \in [1, \infty]$ the equivalence classes of all p-integrable random variables with respect to $\mathbb{P}$ and with $L^0(\Omega, \mathcal{F}, \mathbb{P})$ the space of all random variables on $(\Omega, \mathcal{F})$. Furthermore, let $||\cdot||_p$ be the standard L^p-norm on $L^p(\Omega, \mathcal{F}, \mathbb{P})$. If $X \in L^1(\Omega, \mathcal{F}, \mathbb{P})$ we just say that X is $\mathbb{P}$-*integrable*.

If Q is another probability measure on $(\Omega, \mathcal{F})$ and $\mathcal{G}$ an arbitrary sub-σ-algebra of $\mathcal{F}$, we say that Q is *absolutely continuous* with respect to $\mathbb{P}$ on $\mathcal{G}$ and write $Q \ll_{\mathcal{G}} \mathbb{P}$ if every null set of $\mathbb{P}$ also is a null set of Q., i. e. if for $N \in \mathcal{G}$ it holds

$$\mathbb{P}(N) = 0 \quad \Rightarrow \quad Q(N) = 0.$$

If $Q \ll_{\mathcal{F}} \mathbb{P}$ we just write $Q \ll \mathbb{P}$.

The theorem of Radon–Nikodym provides a characterization of this property.

Theorem A.1. *Q is absolutely continuous with respect to $\mathbb{P}$ on $\mathcal{G}$ if and only if there exists an $(\mathcal{G}, \mathcal{B})$-measurable random variable $L \geq 0$ (i. e. $L \in L^0(\Omega, \mathcal{G}, \mathbb{P})$) such that*

$$\mathbb{E}_Q[X] = \mathbb{E}[X \cdot L], \quad \textit{for all } X \in L^0(\Omega, \mathcal{G}, \mathbb{P}),\ X \geq 0. \tag{A.1}$$

We frequently use the notation

$$\left.\frac{\mathrm{d}Q}{\mathrm{d}\mathbb{P}}\right|_{\mathcal{G}} := L$$

and set

$$\frac{\mathrm{d}Q}{\mathrm{d}\mathbb{P}} := \left.\frac{\mathrm{d}Q}{\mathrm{d}\mathbb{P}}\right|_{\mathcal{F}}.$$

Proof. See Theorem A.8 in Föllmer and Schied (2004). □

As an application, assume that $Q \ll_{\mathcal{G}} \mathbb{P}$ with $L \in L^\infty(\Omega, \mathcal{G}, \mathbb{P})$. As a consequence of the Hölder-inequality, every $\mathbb{P}$-integrable and $(\mathcal{G}, \mathcal{B})$-measurable random variable X is also Q-integrable:

$$\mathbb{E}_Q|X| \leq \mathbb{E}\,|X| \cdot ||L||_\infty < \infty.$$

It follows that (A.1) also holds for general $X \in L^1(\Omega, \mathcal{G}, \mathbb{P})$.

If Q and $\mathbb{P}$ have the same null sets in $\mathcal{G}$, i. e. if $Q \ll_{\mathcal{G}} \mathbb{P}$ and $\mathbb{P} \ll_{\mathcal{G}} Q$, we say that Q and $\mathbb{P}$ are *equivalent* on $\mathcal{G}$ and write $Q \approx_{\mathcal{G}} \mathbb{P}$. This can be characterized as follows:

Proposition A.1. *Let $Q \ll_{\mathcal{G}} \mathbb{P}$. Then*

$$Q \approx_{\mathcal{G}} \mathbb{P} \quad \Leftrightarrow \quad \frac{\mathrm{d}Q}{\mathrm{d}\mathbb{P}}\Big|_{\mathcal{G}} > 0 \ \mathbb{P}\text{-almost surely.}$$

Proof. See Corollary A.9 in Föllmer and Schied (2004). □

This and the following result are important for the investigations in Sections 3.3 and 4.4 and implicitly used there.

Proposition A.2. *Let $Q \ll_{\mathcal{F}} \mathbb{P}$. If $\mathcal{G}$ is an arbitrary sub-σ-algebra of $\mathcal{F}$, then it holds $Q \ll_{\mathcal{G}} \mathbb{P}$ and*

$$\frac{\mathrm{d}Q}{\mathrm{d}\mathbb{P}}\Big|_{\mathcal{G}} = \mathbb{E}\left[\frac{\mathrm{d}Q}{\mathrm{d}\mathbb{P}} \mid \mathcal{G}\right].$$

Proof. See Proposition A.11 in Föllmer and Schied (2004). □

If $(\mathcal{G}_t)_{t=0,1,\dots,T}$ is a filtration on $(\Omega, \mathcal{F})$ and $Q \ll \mathbb{P}$, we can define the so called *density process of Q* with respect to $\mathbb{P}$ denoted by $(L_t^Q)_{t=0,1,\dots,T}$ via

$$L_t^Q := \frac{\mathrm{d}Q}{\mathrm{d}\mathbb{P}}\Big|_{\mathcal{G}_t}, \quad t = 0, 1, \dots, T.$$

By the previous Proposition, $(L_t^Q)_{t=0,1,\dots,T}$ is a non-negative $(\mathcal{G}_t)_{t=0,1,\dots,T}$-martingale. Obviously, every non-negative $(\mathcal{G}_t)_{t=0,1,\dots,T}$-martingale $(M_t)_{t=0,1,\dots,T}$ with $\mathbb{E}\, M_T = 1$ is a density process of Q defined by $\mathrm{d}Q = M_T \mathrm{d}\,\mathbb{P}$.

Finally, let us briefly introduce the *essential supremum (infimum)* of an arbitrary family of random variables or of a random variable respectively. As is discussed in Section A.5 of Föllmer and Schied (2004), it is not very sensible to introduce the pointwise supremum of an uncountable family of random variables. Formally, let J be an arbitrary index set and $(X_j)_{j \in J}$ a family of $(\mathcal{G}, \mathcal{B}$-measurable random variables. In general (if J is uncountable),

$$\big(\sup_{j \in J} X_j\big)(\omega) := \sup_{j \in J} X_j(\omega)$$

will not be $(\mathcal{G}, \mathcal{B})$-measurable or even the right concept. The following result provides a more sensible definition.

Theorem A.2. *There exists a $(\mathcal{G}, \mathcal{B})$-measurable random variable X^* such that*

$$X^* \geq X_j \ \mathbb{P}\text{-almost surely for every } j \in J. \tag{A.2}$$

Furthermore, X^* is unique in the sense that for every other $(\mathcal{G}, \mathcal{B})$-measurable random variable Z fulfilling (A.2) *it holds $Z \geq X^*$ $\mathbb{P}$-almost surely. We set*

$$\operatorname{ess.sup}_{j \in J} X_j := X^*$$

and call X^* the essential supremum *of $(X_j)_{j \in J}$ with respect to $\mathbb{P}$. The* essential supremum *with respect to $\mathbb{P}$ is denoted by*

$$\operatorname*{ess.\,inf}_{j \in J} X_j := -\operatorname*{ess.\,sup}_{j \in J}\{-X_j\}.$$

Proof. See Theorem A.32 in Föllmer and Schied (2004). □

We have seen in Section 1.2 that the essential supremum of a random variable can be used to introduce a (not very reasonable) coherent static risk measure. The former is defined as an essential supremum over a set of constant random variables, namely via

$$\operatorname{ess.\,sup} X := \operatorname{ess.\,sup}\{c \in \mathbb{R} \mid \mathbb{P}(X > c) > 0\}.$$

for a random variable $X \in L^0(\Omega, \mathcal{F}, \mathbb{P})$.

B The Beta distribution

In this chapter we briefly compile some properties of the Beta distribution. Most of the facts can be found in Johnson et al. (1995), for example.

First define the *gamma function* $\Gamma : \mathbb{R}_+^* \to \mathbb{R}_+^*$ via

$$\Gamma(\alpha) := \int_0^\infty t^{\alpha-1} \exp(-t) \mathrm{d}t, \quad \alpha > 0.$$

For integer values it holds

$$\Gamma(n) = (n-1)!, \quad n \in \mathbb{N}.$$

A related object is the *Beta function* $B : \mathbb{R}_+^* \times \mathbb{R}_+^* \to \mathbb{R}*_+$ which can be defined through

$$B(\alpha, \beta) := \frac{\Gamma(\alpha)\Gamma(\beta)}{\Gamma(\alpha+\beta)}, \quad \alpha, \beta > 0.$$

It also has an integral representation of the form

$$B(\alpha, \beta) = \int_0^1 t^{\alpha-1}(1-t)^{\beta-1} \mathrm{d}t, \quad \alpha, \beta > 0.$$

Now, we are ready to introduce the *Beta distribution*. It is absolutely continuous, concentrated on $(0, 1)$ and defined as follows. A random variable X on a given probability space $(\Omega, \mathcal{F}, \mathbb{P})$ follows a Beta distribution with parameters $\alpha, \beta > 0$ if the distribution of X has the Lebesgue density

$$f_X(x) = \frac{x^{\alpha-1}(1-x)^{\beta-1}}{B(\alpha, \beta)} \cdot \mathbb{1}_{(0,1)}(x), \quad x \in \mathbb{R}.$$

We write $X \sim \mathrm{Beta}(\alpha, \beta)$. A special case occurs if $\alpha = \beta = 1$. Then by definition $X \sim \mathcal{U}(0, 1)$. However, the distribution function F_X cannot be calculated explicitly for general parameters.

This distribution arises for example when X_1 and X_2 have Gamma distributions with parameters $\alpha_1, \beta > 0$ and $\alpha_2, \beta > 0$ respectively and by setting

$$X := \frac{X_1}{X_1 + X_2}.$$

Then, X has a Beta distribution with parameters α_1, α_2 for arbitrary $\beta > 0$. In particular, if $\beta = \frac{1}{2}$ and $\alpha_j = \frac{n_j}{2}$ for $n_j \in \mathbb{N}$, $j = 1, 2$ we have that X_j follows a

χ^2-distribution with n_j degrees of freedom for $j = 1, 2$ and $X \sim \text{Beta}(\frac{n_1}{2}, \frac{n_2}{2})$. Finally, let us give some important moments if $X \sim \text{Beta}(\alpha, \beta)$. Denote with $m_X^{(k)}$ the k-th moment of X, $k \in \mathbb{N}$. It holds

$$m_X^{(k)} = \frac{B(\alpha + k, \beta)}{B(\alpha, \beta)} = \frac{\alpha(\alpha+1)\cdots(\alpha+k-1)}{(\alpha+\beta)(\alpha+\beta+1)\cdots(\alpha+\beta+k-1)}. \tag{B.1}$$

Hence,

$$\mathbb{E}\,X = \frac{\alpha}{\alpha+\beta},$$
$$\text{Var}X = \frac{\alpha\beta}{(\alpha+\beta)^2(\alpha+\beta+1)}.$$

Furthermore,

$$\mathbb{E}\left[\frac{1}{X}\right] = \frac{\alpha+\beta-1}{\alpha-1},$$
$$\mathbb{E}\left[\frac{1}{1-X}\right] = \frac{\alpha+\beta-1}{\beta-1}.$$

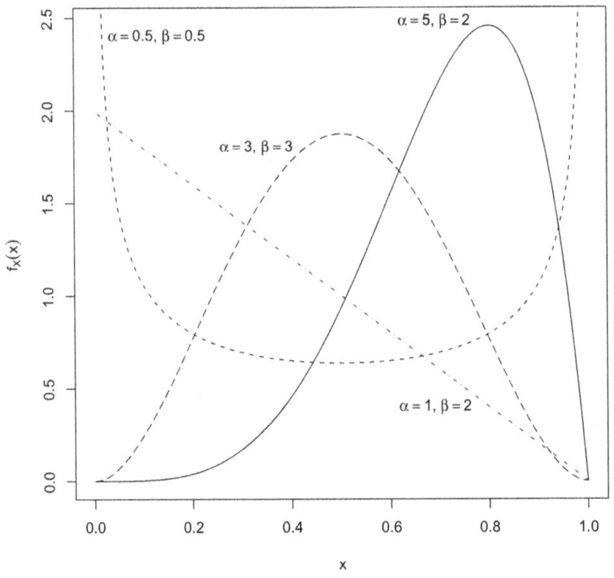

Figure B.1: Densities of $X \sim \text{Beta}(\alpha, \beta)$

Bibliography

ACERBI, C. and NORDIO, C. (2001) Expected Shortfall as a Tool for Financial Risk Management. Technical report, Abaxbank, Milan.

ACERBI, C. and TASCHE, D. (2002) On the coherence of Expected Shortfall. *Journal of Banking and Finance* **26**, 1487–1503.

ARTZNER, P., DELBAEN, F., EBER, J.-M., and HEATH, D. (1999) Coherent measures of risk. *Math. Finance* **9**, 203–228.

ARTZNER, P., DELBAEN, F., EBER, J.-M., HEATH, D., and KU, H. (2004) Coherent multiperiod risk adjusted values and Bellman's principle. Technical report, ETH Zürich.

BARONE, L. (2006) Bruno de Finetti, The Problem of Full-Risk Insurances. *Journal of Investment Management*, **4**.

BÄUERLE, N. and MÜLLER, A. (2006) Stochastic orders and risk measures: consistency and bounds. *Insurance Math. Econom.* **38**, 132–148.

BÄUERLE, N. and MUNDT, A. (2005) Einführung in die Theorie und Praxis von Risikomaßen. In *Risikomanagement*, volume 3 of *Kompetenzzentrum Versicherungswissenschaften GmbH*, 67 – 99. VVW, Karlsruhe.

BELLMAN, R. (1952) On the theory of dynamic programming. *Proc. Nat. Acad. Sci. U. S. A.* **38**, 716–719.

BION-NADAL, J. (2004) Conditional risk measure and robust representation of convex conditional risk measure. Presented at the A.F.F.I. International Conference Cergy-Pontoise 24-26 June 2004.

BION-NADAL, J. (2006a) Dynamic risk measures: time consistency and risk measures from BMO martingales. Technical report, Ecole Polytechnique, Palaiseau Cedex.

BION-NADAL, J. (2006b) Dynamic risk measuring: discrete time in a context of uncertainty and continuous time on a probability space. Technical report, Ecole Polytechnique, Palaiseau Cedex.

BODA, K. and FILAR, J. A. (2006) Time consistent dynamic risk measures. *Math. Methods Oper. Res.* **63**, 169–186.

BURGERT, C. (2005) Darstellungssätze für statische und dynamische Risikomaße mit Anwendungen. Ph.D. thesis, University of Freiburg.

BURGERT, C. and RÜSCHENDORF, L. (2006) Consistent risk measures for portfolio vectors. *Insurance Math. Econom.* **38**, 289–297.

CHERIDITO, P., DELBAEN, F., and KUPPER, M. (2004) Coherent and convex monetary risk measures for bounded càdlàg processes. *Stochastic Process. Appl.* **112**, 1–22.

CHERIDITO, P., DELBAEN, F., and KUPPER, M. (2005) Coherent and convex monetary risk measures for unbounded càdlàg processes. *Finance Stoch.* **9**, 369–387.

CHERIDITO, P., DELBAEN, F., and KUPPER, M. (2006) Dynamic monetary risk measures for bounded discrete-time processes. *Electron. J. Probab.* **11**, no. 3, 57–106 (electronic).

CHERIDITO, P. and KUPPER, M. (2006) Composition of time-consistent dynamic monetary risk measures in discrete time. Technical report, Princeton University, Bendheim Center for Finance.

CUOCO, D., HE, H., and ISSAENKO, S. (2001) Optimal Dynamic Trading Strategies with Risk Limits. FAME Research Paper Series rp60, International Center for Financial Asset Management and Engineering. Available at http://ideas.repec.org/p/fam/rpseri/rp60.html.

CVITANIĆ, J. and KARATZAS, I. (1999) On dynamic measures of risk. *Finance Stoch.* **3**, 451–482.

DELBAEN, F. (2002) Coherent risk measures on general probability spaces. In *Advances in finance and stochastics*, 1–37. Springer, Berlin.

DELBAEN, F. (2006) The structure of m-stable sets and in particular of the set of risk neutral measures. In *In memoriam Paul-André Meyer: Séminaire de Probabilités XXXIX*, volume 1874 of *Lecture Notes in Math.*, 215–258. Springer, Berlin.

DETLEFSEN, K. and SCANDOLO, G. (2005) Conditional and dynamic convex risk measures. *Finance Stoch.* **9**, 539–561.

EPSTEIN, L. G. and SCHNEIDER, M. (2003) Recursive multiple-priors. *J. Econom. Theory* **113**, 1–31.

FAVERO, G. and VARGIOLU, T. (2006) Shortfall risk minimising strategies in the binomial model: characterisation and convergence. *Math. Methods Oper. Res.* **64**, 237–253.

DE FINETTI, B. (1940) Il problema dei "Pieni.". *Giorn. Ist. Ital. Attuari* **11**, 1–88.

FÖLLMER, H. and PENNER, I. (2006) Convex risk measures and the dynamics of their penalty functions. *Stat. & Dec.* **24**, 61–96.

FÖLLMER, H. and SCHIED, A. (2004) *Stochastic finance*, volume 27 of *de Gruyter Studies in Mathematics*. Walter de Gruyter & Co., Berlin, extended edition. An introduction in discrete time.

FRITTELLI, M. and SCANDOLO, G. (2006) Risk measures and capital requirements for processes. *Math. Finance* **16**, 589–612.

HARDY, M. R. and WIRCH, J. L. (2004) The iterated CTE: a dynamic risk measure. *N. Am. Actuar. J.* **8**, 62–75.

VAN HEE, K. M. (1978) *Bayesian control of Markov chains*. Mathematisch Centrum, Amsterdam. Proefschrift, Technische Hogeschool Eindhoven, Eindhoven, March 10, 1978, With a Dutch summary.

HERNÁNDEZ-LERMA, O. and LASSERRE, J. B. (1996) *Discrete-time Markov control processes*, volume 30 of *Applications of Mathematics (New York)*. Springer-Verlag, New York. Basic optimality criteria.

HINDERER, K. (1970) *Foundations of non-stationary dynamic programming with discrete time parameter*. Lecture Notes in Operations Research and Mathematical Systems, Vol. 33. Springer-Verlag, Berlin.

INOUE, A. (2003) On the worst conditional expectation. *J. Math. Anal. Appl.* **286**, 237–247.

IOFFE, A. D. and TIHOMIROV, V. M. (1979) *Theory of extremal problems*, volume 6 of *Studies in Mathematics and its Applications*. North-Holland Publishing Co., Amsterdam. Translated from the Russian by Karol Makowski.

JOBERT, A. and ROGERS, L. C. (2005) Pricing operators and dynamic convex risk measure. Technical report, University of Cambridge.

JOHNSON, N. L., KOTZ, S., and BALAKRISHNAN, N. (1995) *Continuous univariate distributions. Vol. 2* . Wiley, New York, second edition.

JOUINI, E., MEDDEB, M., and TOUZI, N. (2004) Vector-valued coherent risk measures. *Finance Stoch.* **8**, 531–552.

JOUINI, E., SCHACHERMAYER, W., and TOUZI, N. (2006) Law invariant risk measures have the Fatou property. In *Advances in mathematical economics. Vol. 9*, volume 9 of *Adv. Math. Econ.*, 49–71. Springer, Tokyo.

KALLENBERG, O. (2002) *Foundations of modern probability.* Probability and its Applications (New York). Springer-Verlag, New York, second edition.

KUSUOKA, S. (2001) On law invariant coherent risk measures. In *Advances in mathematical economics, Vol. 3*, volume 3 of *Adv. Math. Econ.*, 83–95. Springer, Tokyo.

MARKOWITZ, H. (1952) Portfolio Selection. *Journal of Finance* **7**, 77–91.

MARKOWITZ, H. (2006) De Finetti Scoops Markowitz. *Journal of Investment Management*, **4**.

MCNEIL, A. J., FREY, R., and EMBRECHTS, P. (2005) *Quantitative risk management. Concepts, techniques, and tools.* Princeton Series in Finance. Princeton, NJ: Princeton University Press. xv, 538 p.

PFLUG, G. C. (2006) A value-of-information approach to measuring risk in multiperiod economic activity. *Journal of Banking and Finance* **30**, 695–715.

PFLUG, G. C. and RUSZCZYŃSKI, A. (2005) Measuring risk for income streams. *Comput. Optim. Appl.* **32**, 161–178.

PFLUG, G. C. and RUSZCZYŃSKI, A. (2001) A risk measure for income streams. *Stochastic Programming Electronic Publications* **11**. http://edoc.hu-berlin.de/browsing/speps/.

PIRVU, T. A. (2007) Portfolio optimization under the value-at-risk constraint. *Quant. Finance* **7**, 125–136.

PRESSACCO, F. and SERAFINI, P. (2007) The origins of the mean-variance approach in finance: revisiting de Finetti 65 years later. *Decisions in Economics and Finance* **30**, 19–49. Available at http://ideas.repec.org/a/spr/decfin/v30y2007i1p19-49.html.

PUTERMAN, M. L. (1994) *Markov decision processes: discrete stochastic dynamic programming.* Wiley Series in Probability and Mathematical Statistics: Applied Probability and Statistics. John Wiley & Sons Inc., New York. , A Wiley-Interscience Publication.

RIEDEL, F. (2004) Dynamic coherent risk measures. *Stochastic Process. Appl.* **112**, 185–200.

RIEDER, U. (1975) Bayesian dynamic programming. *Advances in Appl. Probability* **7**, 330–348.

RIEDER, U. (1987) Bayessche Kontrollmodelle. Lecture Notes. University of Ulm.

ROORDA, B. and SCHUMACHER, J. M. (2007) Time consistency conditions for acceptability measures, with an application to Tail Value at Risk. *Insurance Math. Econom.* **40**, 209–230.

ROORDA, B., SCHUMACHER, J. M., and ENGWERDA, J. (2005) Coherent acceptability measures in multiperiod models. *Math. Finance* **15**, 589–612.

RUNGGALDIER, W., TRIVELLATO, B., and VARGIOLU, T. (2002) A Bayesian adaptive control approach to risk management in a binomial model. In *Seminar on Stochastic Analysis, Random Fields and Applications, III (Ascona, 1999)*, volume 52 of *Progr. Probab.*, 243–258. Birkhäuser, Basel.

RUSZCZYŃSKI, A. and SHAPIRO, A. (2006) Conditional risk mappings. *Math. Oper. Res.* **31**, 544–561.

SZEGÖ, G. (2005) Measures of risk. *European J. Oper. Res.* **163**, 5–19.

TASCHE, D. (2002) Expected Shortfall and Beyond. *Journal of Banking and Finance* **26**, 1519–1533.

URYASEV, S. and ROCKAFELLAR, R. T. (2002) Conditional Value-at-Risk for general loss distributions. *Journal of Banking and Finance* **26**, 1443–1471.

WANG, S. S., YOUNG, V. R., and PANJER, H. H. (1997) Axiomatic characterization of insurance prices. *Insurance Math. Econom.* **21**, 173–183.

WANG, T. (1999) A class of dynamic risk measures. Technical report, University of British Columbia.

WEBER, S. (2006) Distribution-invariant risk measures, information, and dynamic consistency. *Math. Finance* **16**, 419–441.

YONG, J. and ZHOU, X. Y. (1999) *Stochastic controls*, volume 43 of *Applications of Mathematics (New York)*. Springer-Verlag, New York. Hamiltonian systems and HJB equations.

After years of hard struggle I finally found and resolved π*...*